George Western Thompson

The Living Forces of the Universe

George Western Thompson

The Living Forces of the Universe

ISBN/EAN: 9783337779269

Printed in Europe, USA, Canada, Australia, Japan

Cover: Foto ©berggeist007 / pixelio.de

More available books at **www.hansebooks.com**

THE

LIVING FORCES OF THE UNIVERSE.

In the *Beginning*, Elohim — the *Almighty Forces* — created. — Gen. i. 1.

I am El Shaddai — *the Almighty God* — Walk before Me and Be Thou Perfect. — Gen. xvii. 1.

In the *Beginning* was the *Logos* — Wisdom — Divine Intellectivity. — Jno. i. 1. Prov. viii. 12–36.

God so loved. — Jno. iii. 16.

Hear, O Israel, Jehovah, our Elohim, one Jehovah. — Deut. vi. 4.

THE TEMPLE AND THE WORSHIPPERS.

KNOW AND GOVERN THYSELF.

Whisper strange secrets in the Whirlwind's ear, for those who sow to the Winds.

BY

GEORGE W. THOMPSON.

PHILADELPHIA:
HOWARD CHALLEN.
New York: Sheldon & Co.; Boston: Lee & Shepherd;
London: Trubner & Co.
1866.

" THE Living Forces of the Universe " presents a new Method and a new Philosophy. Its Method is entirely different from, and more comprehensive and inclusive than, any which has preceded it. It destroys and yet preserves Rationalism by more inclusive and complete elements; and by higher forms and a purer synthesis. The world has been seeking and there has been for a long time a presentiment of this Philosophy or something like it. Dr. Huntington, of Boston, Mass., has given a clear anticipation of it in the tenth sermon of his " Christian Living and Believing." This work fills the exact conditions of his singular and remarkable anticipation as stated in the fourth division quoted below. Those who are accustomed to reflective thought will see, at a glance, that this work is not the result of his conclusion, for that gives no indication of the Method, the processes, the elements, or the facts, nor does he seem to realize, or but faintly, the comprehensive breadth, nor in any precise or philosophical manner the deep foundations, on which the author of the Living Forces rests his system. This work is new in its Method, new in its processes, new in its mode of presentation, new in its doctrines, yet as old as the Thought of Creation, and as fresh and as reconstructive as the moral necessities of the day require.

Dr. Huntington, pp. 188–192, says : —

" Setting aside notions purely Pagan, and keeping in the line of the nominal belief in one God, there are three distinctly marked stages in the progress of opinion about the natural world, with a fourth to come.

" The first of these is where the natural world is regarded as divine only as to what appears to be extraordinary or exceptional in it. Thunders, tempests, earthquakes, eclipses, famines, pestilences, are thought to betray a divine presence. Or, in human affairs, sudden accidents, unexpected deliverances, strange coincidences. God is a God of occasional interference, not of constant regulation and animation. Not all our daily affairs and the regular processes of creation are subject to his watchfulness, and charged with his indwelling spirit;

but nature is liable to arbitrary visitations from without. The religious sentiment feeds on the marvellous. There is a piety of surprises
and alarms, — intermittent, spasmodic. God is not in the order of nature, its laws, its silent, beneficent growths and noiseless motions, but
in its loud jars and grotesque anomalies. You will bear much there
of special providences: it is not Providence at all, but intrusion, im-
provisation, perturbation. Of course this will be a God of violence
and of terror. And the name of this first view will be Superstition.
The supernatural is, then, strange, frightful.

"The second is exactly opposite to this. It is where the attention is
turned wholly to the law-side of nature, and does not see that there is
a personal will acting freely anywhere within nature or about it. It
is so bent on getting rid of exceptions that it forgets the Maker. It
mistakes uniformity for self-acting mechanics. Virtually it denies the
spiritual world, with all its nobler, varied and glorified forms of life.
There are men so absorbed in the regular processes of the universe as
to be insensible both to its Original and to its holy object. Prudence
is substituted for piety. The nearest approach to penitence is regret
for a miscalculation. Self-reliance is put for devout trust: a little
knowledge, which vanishes away, for faith and hope and charity,
which abide. The future is all dark, without promise or resurrection.
The name of this is skepticism. The supernatural is denied.

"The third, which is unquestionably a great advance on the other
two, is where God is believed to be over both the natural and the spiritual world, but only in the spiritual. These two worlds are driven
wide apart. Thus the only religious purpose answered by nature is
to furnish a convenient supply of figures and illustrations for religious
discourse. In those who have a lively admiration for external beauty
there will grow up a sort of fanciful, poetical, sentimental piety: in
those who distrust and despise the material world, asceticism. Christianity and creation are sundered, though God joined them together.
It is a kind of half-belief. The supernatural is essentially unreal: and
the evidence of miracle, where it is introduced into theology, has a
materialistic cast, as if the high and self-attesting truths of Christianity and the soul were actually *dependent* on proofs addressed to the
senses.

"But there is a fourth condition, — or will be yet, — *where the natural
and the spiritual are seen and felt to be parts of one plan, under one
Creator.* The laws of the one are recognized to be exactly harmonious, nay, identical with the laws of the other. There is not only a
resemblance, but a correspondence: the things of nature being found
to be the things of the spirit of man, good and evil; and all the things
of nature having their counterpart in the spiritual world, whether life
or death, health or disease, clouds or sunshine, serpents or doves.

Christ's instructions are full of these things; and they are not acci-
dental comparisons, but are meant to bring God's works together into
the closest unity. So says the Apostle Paul in a passage which com-
mentators have only partially and superficially comprehended: 'The
invisible things of Him, from the creation of the world, are clearly
seen, being understood by the things that are made.' He is the God of
the insect as much as of the archangel. In the original design of the
Creative mind, each was meant for the other, — everything in nature,
great or small, star or starfish, to meet and answer to something in
man. This at present may be Christian mysticism. But it will be
Christian faith. All the strong tendencies of true science, as well as
of Revelation, are bearing in this direction. They tell us that when
God formed the lowest living creature, already man, with brain and
heart and immortality, was in his thought. In every department of
knowledge and thought, unity is the reigning idea. All interdepend;
all belong to each other; all serve each other. And this is the Christian doctrine. Revelation is to find each of its great practical truths
confirmed in the universe. The sovereignty of God; his personal and
free presence to every part and particle; the disorder of sin or disobedience to law; the remedy for that, or reconciliation; the necessity of
a second or spiritual birth to restore and complete the natural man, —
have dim types in nature. And, above all, — what now concerns us
most, — there is hinted the reality of a revelation of what is unseen
and eternal, through appropriate and pre-adapted forms. that are seen
and temporal, in connection with the ministry of the Son of God and
Son of Man, as a mediator belonging both to earth and heaven, or
rather as having both these belonging to him. In this view, the Christian miracles become not only credible, but what we should have a
right to expect; such breakings through of the spiritual upon the ordinary world as a mediator's ministry would probably bring with it,
and the only rational explanation of the beginnings of Christian history.

"As to this Revelation, then, the first of these four views I have
mentioned — superstition — is ignorant of it; the second — skepticism
— rejects it; the third misinterprets it; the fourth — faith — finds it
full of blessed meaning, and brimming at every point with a heavenly
inspiration."

THE CLUE TO THE CONCILIATION OF REASON AND RELIGION, AND THE INSTAURATION OF LOVE.

———◆———

IN passing from the Being to the Nature of God we are compelled to reason from Ourselves ; for from ourselves, from our own Higher Nature, a pathway is found to the Highest Nature of all. — *Christ of History.*

So must the laws and phenomena of the human mind be correctly analyzed and clearly defined, in order to obtain a clear Insight into the Intellectual System of the Universe. And just in proportion as the clouds and darkness hanging over the phenomena of our own Minds are made to disappear, will the Intellectual [and Moral] system of the world which God has "set in our hearts" become more distinct and beautiful in *their* proportions. — *Bledsoe's Theodicy,* In. vi.

And God said, Let Us make Man in Our Image after Our Likeness ; and let them have dominion. — *Gen.* i. 26.

He, the Logos, the Word, was in the World, and the World was made by Him. — *Jno.* i. 10.

I will pray the Father, and he shall give you *another* Comforter, that he may abide with you forever — the *Spirit of Truth.* — *Jno.* xiv. 16, 17.

Let Us make man in our Image (εἰκών, LXX. אֶלֶם) after our Likeness (ὁμοίωσις, LXX. דְמוּת). The Alexandrians taught that the εἰκών, *image,* was something in which men were created, being *common* to all and continuing to man after the Fall as before, Gen. ix. 6 ; while the ὁμοίωσις, the *likeness,* was something *toward* which man was created that *he might strive after and attain it ; Origen, Princ.* iii. 6 : *Imaginis* dignitatem in prima conditione percepit, *Similitudinis,* vero, perfectio in consummatione servata est ; cf. *in Joan,* tom. xx. 20 ; *The dignity of the Image is perceived in the first state of man, but the perfection of the Likeness is attained in the Consummation.* And the Schoolmen : Imago secundum cognitionem veritatis, Similitudo secundum amorem virtutis ; *The Image is according to the Cognition of Truth, the Likeness is according to the Love of Virtue. . . . .* We may expect to find mysteries there; prophetic intimations of truths which it might require ages and ages to develop. And without attempting to draw a very strict line between εἰκών, image, and ὁμοίωσις, likeness, or the Hebrew originals, I think we may be bold to say, that the whole history of man, not only in his original creation, but also in his after-restoration and reconstitution in the Son, is significantly wrapt up in this double statement; which is double for this very cause — that the Divine Mind did not stop at the contemplation of his first creation, but looked on him as "renewed in knowledge after the Image of Him that created him," Col. iii. 10; because it knew that only as partaker of *this* double Benefit would he *attain* this true end for which he was made. — *Trench's Syn. N. T.* § xv.

You say, Who can believe that in one God there are three Persons? Observing that by the three Persons we do not understand three individuals, but three *distinct relations* subsisting in one nature, I ask in my turn, Is this mystery more incomprehensible than the eternity of God? — *Protestantism and Infidelity*, c. iv. § 2; *F. X. Weninger, D. D., Miss. of the Society of Jesus.*

In the Beginning God *created*. — *Gen.* i. 1.

And the Lord God said, Behold, the man has become as one of Us to *know* Good and Evil. *Gen.* iii. 22, 25. The New Man which is renewed in knowledge after the Image of Him that created him. — *Col.* iii. 10. Christ Jesus who, of God, is made unto us Wisdom and Righteousness. — 1 *Cor.* i. 30. God is *Love;* and he that dwelleth in Love dwelleth in God and God in him. 1 *Jno.* iv. 16. Love is the fulfilling of the Law. — *Rom.* xiii. 10.

Stir up the gift of God which is in Thee . . . of *Power* and of *Love* and of a *Sound Mind.* — 2 *Tim.* i. 6, 7.

And may the God of Peace sanctify you in all things that your whole *Spirit*, Πνεῦμα, and *Soul*, Ψυχή, and *Body*, Σῶμα, may be preserved blameless in the coming of our Lord Jesus Christ. — 1 *Thess.* v. 23. Heb. iv. 12.

To as many as received Him — the Logos — Wisdom — gave he power to become the Sons of God. — *Jno.* i. 12.

God is a Spirit, and they that worship Him must worship Him in Spirit and in Truth. — *Jno.* iv. 24.

God hath revealed them unto us by his Spirit, for the Spirit searcheth all things, yea, the deep things of God. — 1 *Cor.* ii. 10.

What man knoweth the things of a man, save the Spirit of man which is in him? — 1 *Cor.* ii. 11.

THERE is herein a union of the philosophic, dogmatic, and didactic styles, and these frequently in
the same section and at times in the same sentence.
It is a philosophy of dogmas. It offers solutions for
many of the Contradictories which have appeared
in all the philosophic schools, from the earliest to
the present time ; and it tenders Conciliations of
the Dogmas of all religions, creeds, and superstitions
which are capable of being resolved into a consentaneous, harmonious, and progressive system for the
culture and advancement of Humanity. It sets
forth a System of the Universe predicated on the
demonstration that God is an All-Mighty, All-Wise,
and All-Loving Being. Its Method is a novelty,
yet is as old as the Teachings " about all Galilee."
The philosophy is legitimated in the Method and
its processes, and no dogma is stated without its
demonstration and its appropriate and syntactic
designation in the universal system. Theological
Dogmas which have controlled the actions and
guided the conduct of the most highly gifted per-
sonages, through many centuries of the past, when
legitimated in broad and deep foundational processes
which include all facts and all elements of the life-

movement of the Cosmos, acquire a rich value for mankind.

Truth and Dogma must be, in their didactic uses, positively affirmed, for Truth is the final concurrence of human thinking with the processes of nature and life, unfolding in subordination to their primary laws and ultimate ends. Truth should be didactically enforced, or it is empty, useless, and vain. Independent propositions may remain, yet, incapable of conciliation. It is a great weakness, and it may be a great wickedness to oppose such independent propositions one against the other, when each of them, by itself, is capable of reasonable or satisfactory demonstration. Take each demonstration of the true for the truth so far, and perhaps in time, as by the resolution of many previous doubts in the search after Truth, the conciliation may come.

Those whose minds are filled with a love of Truth and a desire to reach forth personally to God as a Personality and to return to the earth for the discharge of serene and solemn duties, can, with a clear and open mind and loving heart, move through the Temple of the Universe, and see the many Worshippers and Workmen in their different stages of Advancement, from the infancy of tribes through the gradations of life and history, and catch some systematic view of life, and more or less intelligibly and lovingly discharge their duties in the unfolding series and cycles, on and still on, or else sink down and still down through the long ages, and it may be through the multitudinous worlds. Those who are prepared to follow the Clue of the movement, already placed in their hands, yet firmly legitimated in the processes herein instituted, and in some learning of the past and the present, with a heart full of hope for the future amid the ruin around and the mutterings of threatenings in every part of the world, will see from one standpoint that it is a Philosophy springing from an elementarily Foundational Religion, and from another, that it is a Religion springing from an elementarily Foundational Philosophy. It is the Coördination of both.

If the language is somewhat new, it is necessarily so. A new philosophy cannot be transparent in old expressions, and old dogmas which have hardened into forms unsuited to the living wants of the age must, in their legitimation, seek new exponent terms for their underlying truths. A New Life must appear in a New Form, albeit that Form, as in all life, may be composed of the disintegrated elements of older *forms*. Greek life could not have been manifested in the Egyptian forms, nor the Roman in those of the Greek; and modern thought and feeling could not be realized or conveyed in the forms of the Middle Ages, and German, French, and English modes of expression, confused and perplexed with their various meanings and diversified applications, growing out of precedent conditions, are unsuited to convey a life of Love and Thought and Actuation which may embrace and fuse and comprehend the Whole. The language, adaptive to a life of *dis*-envelopment with its manifold activities, must differ from that of its germinal *en*-velopment, although it must embrace it.

There are no novelties here, except in the method, yet all is new. The foundations of what is written have been given to me by a solemn instruction through strange providences and sad vicissitudes, yet maintaining through all an earnest and open spirit of inquiry and of receptivity. They ask no faith or belief except on their demonstrations of Truth. We are in the Movement of a great Prolepsis ; and under these strange and instructive Providences, obedient and trustful, I would discharge my solemn duties, and, amid the madness of the distempered times, would contribute to restore Peace and Charity to men's bosoms, or to reanimate a new race rising into manhood. At such a time and to such fresh minds I would commit the gift which has been given to me — the Life, the clearer and purer Life of Old Truths, from a serener standpoint, vouchsafed nearer to the elementary, the Essential Foundations.

CONTENTS.

BOOK FIRST.

BOOK SECOND.

THE PHENOMENOLOGY OF NATURE AND LIFE, OR NATURE AND LIFE AS REPRESENTATIVE OF AND PHENOMENAL OF ONTOLOGIC BEING.

BOOK THIRD.

ONTOLOGY AND PHENOMENA IN THEIR PROLEPSIS.

BOOK FIRST.

DEFINITIONS AND FOUNDATIONAL THOUGHTS.

THE authorities adduced are the most orthodox and exclusive in Religion, the most approved in the respective Sciences, and the most conservative in Philosophy.

In the References, throughout, the Numerals and Figures, as follow, will refer to the Books, Chapters, and Sections of this Work ; thus : I. iii. 15, is B. I. c. iii. § 15 ; and II. vii. 31, is B. II. c. vii. § 31, &c.

Well knowing the tendency to degradation in vulgar, rude, animalistic, and human Imaginates, and in a language which corresponds with and embodies them, and conscious of the necessity of pure Ideas and of the proper dignity and exaltation of expression which should accompany them, the latter have been adopted, and rather than lower these to the standard of a life which needs all elements of purification and elevation, a Glossary of a few words, not current among general readers, is added with the hope that the work will be more widely useful and acceptable. B. I. c. v. § 26 ; c. vi. § 40.

GLOSSARY.

Accidence. A property or quality which may or may not be superadded to a thing, or to the condition of a thing. That which may belong to the thing, but is not essential to it.

Adumbrate. To shadow forth from an inner light.

Æsthetics. The science and culture of matters and forms of taste.

Afferent. Nerves which carry inwardly or from a ganglion of originating force communicating sensations or nervous power.

Afflatus. An inblowing, inbreathing of spiritual life or power.

Antithesis. An opposition or reverse power of action or mode of thought.

A posteriori. I. iv. 4.

A priori. I. iv. 3.

Autocthon. One who rises or springs from the ground which he inhabits.

Automatic. I. i. 32.

Autonomy. I. i. 27.

Autopsy. I. i. 31.

Axis-cylinder. The central substance of a nerve-fibre.

Caudate. A ganglion or nervous centre with a tail-like pro longation.

Causal. Containing in itself the elements or an element of causation.

Centripetate. Tending or drawing to the centre.

Charlatanerie. Science or art prostituted to fraud or quackery.

Circumfer. To flow or bear around and touch at every point.

Commissural Fibre. The rudimentary brain of the lower orders of animal life.

Compages. A system of structure of many parts united.

Complement. That fulness of quantity and quality — the content, which makes a thing or subject complete.

b

Concrete. Is opposed to the mere abstract idea — the empty form. It is that union or concentration of all which is necessary to constitute the substance of the thing and give content to the Idea.

Congeries. A collection of several or many organs to complete a body or aggregate, and make it a more or less perfect whole.

Consubstantial. Substances which in cause and effect produce similar or identical results.

Contingency. I. i. 12.

Contour. The outline that defines a figure.

Convolute. The brain presents the appearance, somewhat, of leaves in the bud or petals in an opening flower. Each of these convolutes may well be supposed to have its office or function.

Coördinations. I. i. 4.

Correlations. I. i. 6.

Cosmos. Cosmical. The system of all the systems of the heavenly bodies. — *Cosmical matter*, the various kinds of substances and forces of which the stars, comets, planets, &c., and all they contain as such, were formed.

Diaphanous. Transmitting light, intelligence from beyond.

Differentiation. I. i. 25, 29.

Discrete. That which is separate in virtue of its own distinctive nature.

Dynamic. I. i. 25.

Ectype. The impression — the thing made from the type.

Efferent. The nerves which carry out action or motion. Where the afferent nerve enters the axis-cylinder of the efferent nerve it may well be supposed, here, that the impulsion given to the afferent is continued by the efferent without break or modification of the original impulsion. I. i. 30; vi. 11. Where the communication is not thus direct, but the afferent force is distributed to a number of ganglia or nerve-cells, other modifying forces may be called into action. And where it is communicated to the Autopsic Self, it may act by the independent organism placed under its determinate control.

Egressus. That movement by which the Self goes out of itself as it were, into nature and life.

Emanation. The movement towards the formation of the Cosmos, which implies that the transforming forces are mere developing spontaneities.

Embryology. The forming rudiments of things acting under natural causes.

Endemic. Special to a people or tribe in a locality, as a physical or mental disease.

Entity. Something having a discrete, positive existence.

Exacerbation. Passions, affections, sentiments, or conditions of organs in which they are not merely excited, but unnaturally inflamed and morally or physically diseased, so as to produce diseased and malignant action.

Fascicle. A bundle. A number of observed facts or phenomena referred to the thing in which they inhere or from which they make their appearance.

Filamentary. In threads or thread-like connections.

· *Functionalize.* The inweaving of forces and giving of organisms to each vegetal and animal thing, by which it performs, executes, that which is proper to each organ. And as many organs have particular functions, each is severally functionalized. Properly, all created things and forces are functionalized.

Fusiform Spindle-shaped. Thick and tapering to each end. A shape of a ganglion.

Ganglion. A mass of nervous matter forming a centre from which nervous fibres radiate; and is of various forms, as caudate, fusiform, stellate, or spheroidal; and when one fasciale so radiates it is called unipolar, or two, it is called bipolar. See Efferent. I. iii.

Genesis. Creating. Producing.

Geotic. All terrestrial causes which act upon and modify the human system in any locality or place.

Germinal. An improvement and perfecting by growth and culture.

Gradus. The ascent by which the Self unfolds into higher and still higher forms of life.

Hemiplegia. A palsy that affects one half of the body.

Homogeneous. The same in every part and in the whole.

Homologue. A lower form of organic structure, having a likeness of form and function to a higher form of organic structure.

Hypnotism. Somnambulism, or the artificial state of mesmeric sleep.

Hypostatic. Hypostase is discrete essence or substance or power.

Idea. Ideate. Ideation. I. vi. 40 ; v. 16.

Identity. Discrete essence or substance. I. ii. 14 ; iii. 18.

Ideological. Those mental processes confined exclusively to and in the Intellectivity.

Immanent and *Permanent* are to a certain extent antithetical in philosophy. *Immanent* is applied to forces which are *made* by the divine constitution of things intrinsic, inherent, in virtue of which they *continue* to act. *Permanent* is the continuous and immediate presence of Deity *moving* the Forces. The latter is Pantheism.

Implicate. The inwoven and correlated system of the Cosmos.

Inosculate. United by opposition or contact so as to communicate from one to the other.

Insistent. I. i. 85 ; vi. 7.

In situ. Fixed in place.

Instauration. The restoration to moral order.

Intelligential. Having intelligence inwoven in it, but not consciously intelligent.

Inter. In composition of words, is — between ; *inter*currence.

Internuncial. I. i. 80 ; vi. 11.

Introspection. The power of looking within one's self and separating the passions, affections, and intellections in their kinds and forms of action.

Intuition. Intellectual insight by and of the simple cognitive power of the Self.

Isothermal. The lines around the earth having equal degrees of heat.

Libration. The power by which one passion or affection is brought to affect or balance or counterbalance another.

Locum tenens. Place of occupancy.

Macrocosm. The Cosmos in its living forces, dependences, and working correlations ; and the word has been used always as having some resemblance in the microcosm or the organization of man.

Menstruums. Substances as recipients of forces and from which forces can be resolved and separated.

Mobilized. Moving with an aptitude of motion or action in itself.

Modality. The quality of being modal or giving form.

Momenta. The movement-power.

Mow. To grimace ; distort the face.

Mysticism. I. iv. 2, 12.

Normalate. I. i. 33.

Notionalize. I. v. 28.

Noumenon. Those substances or subsistences which underlie actual phenomena. It is the thing or force which is notion-alized. The discrete identity.

Objectiv-facient. Setting over in independent or quasi-independent entity. I. vi. 10.

Ontology. The science of foundational causes. That from which primal causations are initiated.

Opinion. I. i.; v. 29.

Orgasm. A condition of excitement and turgescence of an organ, usually applied to the venereal passion. It is the particular functionalization which gives excitement and tendency to action in any and each passion, appetite, affection, desire, &c. I. iii. 11; v. 11.

Orma. I. v. 33.

Oscillate. Swaying between and in virtue of its own tendencies.

Paradox. That which is true and proper in one condition of things, and false and improper in another.

Permanent. See *Immanent.*

Permeate. To pass through without rupture or displacement of parts and fill the interstices with that which permeates.

Persiflage. Bantering talk or trifling style of treating a subject. The witticisms of the buffoon.

Perspicacity. Acuteness of mental discernment.

Perspicience. The act of acute mental discernment.

Pervade. To pass or spread through the whole extent of a thing and into every minute part.

Plastic. I. i. 42.

Predicate. A *predicate of fact* is that which is truly affirmed of something, and without which it is not that which it is in its true entireness as a complete whole. A *predicate of language* is something affirmed of a thing which it may or may not possess without destroying its identity as a complete whole.

Primordial. Original forces. First movement of forces.

Prolepsis. I. i. 42.

Psychology. The Science of the Soul. The art of introspective self-analysis.

Psytations. I. v. 10–14.

Quasi. That which is so, but not wholly so.

Racemate. To form and grow into clusters.

Rationalism. Philosophizing exclusively in ideological processes

Redactive. Giving *form*. I. i. 20, 23, 40.

Redintegrate. To renew ; to restore to a full or perfect state.

Reflex. Reflex act. I. vi. 6.

Relation. I. i. 7 ; vi. 23.

Repercussing. Where cause which produces action in one organism is conducted to another, and that is put into action, this last action is a repercussion. It will be seen by close introspection and observation that this is the mode of action in animal and most of human psychical action. It is the unbroken movement of the original cause of movement. The self-conscious Self can frequently break up this movement and prevent the repercussion ; and it can from on the other side send down forces by which it will play off one passion or affection against another, or restrain action.

Solidarity. I. i. 84.

Somatic. Relating exclusively to the body-life.

Spheroidal. A flattened or prolonged sphere.

Sporadic. Scattered, disjointed, not systematic.

Stabilitation. I. i. 12.

Static. Resting in place by mere weight. Attractive force.

Stellate. Star-shaped.

Sub modo. In some special limited form.

Sustension. Preservation of identity in substance or form.

Synchonic. Existing at the same time.

Syntax. Union of things in connected system or order.

Teleologic. The End which is foreseen and involved in the processes from the beginning and consummated in the End.

Transcendental. I. i. 41, 42, 43.

Triplicate. Three diversities which are necessary to make a whole, and produce the proper action of each and of the whole.

Ultroneous. That determinate action of the Self in which the triplicate powers are concerned, though it may be in different degrees.

Vesicular. Having small membraneous cavities.

Zoic. The immortal life as distinguished against the biotic — body-life and the psychic life — a distinction palpable in the New Testament in various passages.

BOOK FIRST.

FOUNDATION STONES.

—◆—

CHAPTER FIRST.

FUNDAMENTAL DEFINITIONS AND THOUGHTS.

MAN is a Triplicity in Unity. He is Body, Soul, and Spirit. 1 Thess. v. 23 ; Heb. iv. 12. In Body, he is a congeries of organisms bound together in an inclusive organization, which gives him his *form* as man, while each organ has its appropriate *form* for the content and exercise of its special function, and thus for its operation and manifestation. These organs are correlated, in various manners, to his special soul-organisms, each. with their organic powers of functionalization. In Soul, he is a congeries of organisms, in which is infolded and inwoven the soul-forces of his psychical nature, as cunning in the fox, ferocity in the tiger, secretiveness or theft in the crow, &c. &c. These soul-organisms are correlated to various organisms of the body for outward manifestation and connection with external nature, for receiving influences from without, and for transmitting outwardly to nature and into life ; and they are correlated for transmitting these influences inwardly to the Spirit, and from the spirit, by its autopsic powers, to the various organisms, both of these soul-organisms, and through them to the corporeal organisms by which it acts on nature and in life. In the Spirit are found correlations of certain forces, which, to use for the present the language of philosophy and theology, are known as will, by which it acts, objectifies itself, — of intellect, by which it thinks, selects motives, forms plans, devises ways and means, and determines its places and times of action, — and the feelings, which are of its affectional nature. This is man's spiritual Triplicity in Unity. To unfold these elements as foundational in the spiritual nature of man and as co-essential in Being — God, and to catch the filaments of their correlations, require a new Method in philosophy. Such new method is proposed, and a legitimation and conciliation of many of the contradictories, which have confused and confounded philosophy, and made religious dogmas odious and sometimes contemptible, is the aim, and, it is believed, is the success of the Redactive system of the universe, built upon the method of Intusception, — a going into the organic functions of nature and life, and reënlifing them, and comprehending them in virtue of these threefold spiritual powers of the self, which are the image and to be made the likeness of those powers by which all things were made that are made, by which they are sustained and operated, and through which they are to be restored to order and consequent harmony.

1. INTUSCEPTION, abbreviated from Intussusception, is a term to which much importance is attached. It is an old word with a new meaning. It is, from its reverend use by one of the most learned writers of English philosophy, that the word is now borrowed and used in a sense, analogously, derivative from his meaning. In his system it meant, that accordance or agreement, that transparency of the body from the light or influence

of the spirit, and the nature within, by which the complete appearance, the frame, contour, actions and expressions of outlook, as well as of speech and conduct, is the direct representative, the intelligible picture of the internal man. Man becomes the diaphanous ectype of the inner spiritual self, as he is moulded and moulds his surrounding organisms, from instant to instant, in their animalistic propensities, their human desires and purposes, and in his higher spiritual manifestations of autopsic *willing, intellectualizing*, and *loving.* This is stating it more fully than he had intellectualized his own conception. The person is thus a symbol, moulding to represent the beast, the man, the viscous spirit or the holy influence within, as the one or the other temporarily or permanently prevails. In looking on a well-known and thoroughly comprehended neighbor, we see in his very conformation the prevalence of animal passions, obstinate will, mere sentimalities, high moral qualities, the habitudes of a purely natural temperament, or the culture and grace of life as they shall engrave their effects. These are so presented in, and consociated with his appearance, and so uniformly characterize men of the same respective qualities, that the one suggests the other with a sense of conformity and fitness; and they furnish model illustrations for the works of artists, the copies of actors, and the judgments of general character. It is seen in the cunning of the fox, the ferocity of the tiger, the boldness of the lion, and indicates the qualities of many of the varieties of dogs, and is appreciated in all distinctive knowledge of the animal races. In this view, to a higher spiritual observer, and absolutely in the sight of an omniscient intelligence, the whole form and structure of the man becomes, as from instant to instant he moulds himself, consciously or unconsciously,

the exact ectype and adumbration of the inner man.
II. iv. Thus man suffused, enlifed by the elements of
his own intelligibility, is read by the intelligences around
and above him : so man reads animal natures, as well as
his neighbors, in their forms and habits, and thus catches
the correspondence between spirit and bodily form, between functions and organisms, — between creative powers and redactive forms. The instinct of each animal is
inwoven in its special organism ; so the varied instincts
of the human race have each its representative organ, as
that race has its psychical organism placed in juxtaposi-
tion with the instinctive organisms for correspondence
with the autopsic self, and to end in the subjugation of
this spiritual self, or in its control and mastery over the
animal and the man. In the full and matured observation of life this outward similitude in men and animals
is seen to involve the fact of a corresponding similitude
or identity of inner organization up to certain points
of instinctive and psychical powers. The cunning in the
fox or other animal is but cunning in man ; the sagacity
in the elephant is sagacity in man; the song of the mocking-bird is music and poetry in man ; so constructiveness
in the bee and the beaver have their respective organisms for inherence and manifestation, and are homologous
to similar but more complicate organisms in man, yet
connected with other and higher endowments in the
human races. Now observe that in all organic as well
as crystalline nature there are certain hidden and unap-
preciable forces, except in their intelligential and intelligible effects, which produce and mould into form and
qualities the separate organism and subordinate crystallizations of each thing, and that there is to each thing a
correlation of forces which subordinates these organisms,
and their differentiate forces to the respective *form* of

each thing. To intuscept and know nature and life, the self must do more than cognize these redactive forms, for it must enter into the forms and reënlife them, and catch the forces which build and mould them, and in some manner *live* the very functions which operate' in them. Thus man intuscepts the cunning, the ferocity, the sagacity, the constructiveness of animals, and the intelligible forces of the crystalline, the vegetal, and the animal kingdoms.

"A discussion of the problem of human Sociology would, therefore, only be completed after a study of the same problem in the entire animal series, a task requiring varied and profound knowledge of natural history and comparative anatomy. . . . . The social problems presented to us by animals are a fitting introduction to the social problems of man." — *Draper's Phys.* 603.

"And this strange fact of the progress of the human brain is assuredly a fact none the less worth looking at from the circumstance that infidelity has looked at it first. On no principle, recognizable in right reason, can it be urged in support of the development hypothesis; it is a fact of fœtal development and of that only. But it would be well should it lead our metaphysicians to inquire whether they have not been rendering their science too insulated and exclusive; and whether the mind that *works by a brain*, thus 'fearfully and wonderfully made,' ought not to be viewed rather in connection with all animated nature, especially as we find nature exemplified in the various vertebral forms, than as a thing fundamentally abstract and distinct. The brain, built up of all the types of brain, may be the organ of a mind compounded (! ?), if I may so express myself, of all the varieties of *mind*," — he should have said — of Functionalizations. — *Foot-Prints*, by Hugh Miller, ch. xx. —

The truth and the confusion of these remarks will appear.

" If we investigate the condition of the various orders of vertebrate animals, which alone admit of a comparison with our own species, we find on the one hand great differences among them with regard to both their physical and mental faculties; and on the other hand a not less marked difference as to the structure of their brain. In all of them the brain has a central organ, which is the continuation of the spinal cord, and to which anatomists have given the name of the Medulla Oblongata. In connection with this there are other bodies placed in pairs, of a small size and simple structure, in the lowest species of fish, becoming *gradually larger and more complex* as we trace them through the other classes, until they reach their greatest degree of development in man himself. That each of these bodies has its peculiar functions, there cannot, I apprehend, be the smallest doubt; and it is indeed sufficiently probable, that each of them is not a single organ, but a congeries of organs, having distinct and separate uses." — Sir B. Brodie's *Mind & Matter*, 43.

" Wherever there is organization, even under the simplest form, there we are sure to find instinctive action, more or less in amount, destined to give the appropriate effect to it. This is true throughout every part of the animal series, from man and the quadrumana, down to the lowest form of infusorial life. When we consider how vast this scale is — crowded with more than a hundred thousand recognized species, exclusively of those which fossil geology has disclosed to us — we may be well amazed by this profuse variety of instinctive action ; as multiplied in kind as are the organic forms with which it is associated, and all derived from one common power." Sir Henry Holland, cited *id.* 178.

Carpenter, II. P. § 568, says on this subject: " Hence the cerebral hemispheres of man include an amount of nervous matter which is four times that of all the rest of the cranio-spinal mass, more than eight times that of the cerebellum, thirteen times that of the medulla oblongata, &c., and twenty-four times that of the spinal cord. The average weight of the whole encephálon, in proportion to that of the body in man, taking the average of a great number of observations, is 1 to 36. This is a much larger proportion than that which obtains in most other animals; thus the average of mammalia is stated by M. Leuret to be 1 to 186; that of birds, 1 to 212; that of reptiles, 1 to 1321; and that of fishes, 1 to 5668. It is interesting to remark, in reference to these estimates, that the encephalic prolongation of the Medulla Oblongata in man (being about one sixteenth of the weight of the whole encephalon) is alone twice as heavy in proportion to his body as the entire Encephalon of reptiles, and ten times as heavy as that of fish."

And Dalton, H. Phys. 364, says: " The number and relative size of these ganglia, in different kinds of animals, depend upon the perfection of the bodily organization in general, and more especially on that of the intelligence and the special senses."

2. Nature is intusceptible. Intusception is the conscious ingoing, the discriminate injection of the self into the forms and processes of nature and life. It is herein used to designate the whole of the processes by which the self gains knowledge of the whole of the *moving* forces which furnish forth nature and life. It is by injecting the self — by interpenetrating, transfusing — by going forth from itself, as it were, in some or all of its psychical movements of willing, intellectualizing, or loving and infusing, interfusing, circumfusing with its consciousness — the

conscious self, with these very powers of the self, that
which is made the object of knowledge, that man obtains
the diverse knowledge of nature and of life. It is the
self in its own appropriate functions of specific actions
which moves forth and acquires this knowledge; that is,
without the capacity to *will*, and without actually willing,
it cannot understand aught of will in others; without
intellectivity and without actually intellectualizing, it
cannot comprehend intelligibility or intelligence in aught
else; so of the affections. This necessity for Intusception holds equally whether the object of cognition is
without the mind, that is, is objective or is within, and of
the proper self, that is, is subjective. The method of
the process will come up more clearly when applied to a
subjectivity in and of the self. That which is purely
subjective cannot be cognized in its simple subjectivity,
but must be made objective to the self by reflection,
— by a reflex act of the mind reproducing the act,
intellection or affection, for the purpose of being desig-
nately examined; that is, it must be reproduced in the
consciousness by a voluntary process of reënlifement for
its examination, or it must be caught in some of its
effects upon its corresponding muscles or viscera before
its effects have faded out and be thus intuscepted — re-
enlifed. So in objectivities; they are only apprehended
and comprehended, each of its kind, in proportion as the
self infuses and circumferes or enters into them in their
construction or processes of production or action. Everything, therefore, must in a certain definite sense be a *subject* of knowledge before it can properly be an *object* of
knowledge — of cognition. The first step towards cognition is the reception of sensations in, within the self. It
is the modification of the self by the object cognized, and
it is the self going to, into, and around the object, and

taking note of all the points and qualities of the object, and supplying, in all cases, from itself — from its own animalistic or psychical or spiritual nature or the whole of them, the *intelligible* elements in the object observed. In the currents of different cognitions, by their frequent repetitions, are given the uniformity and verity of the modifying objectivities and of the psychical and spiritual elements concerned in the conscious operations, thus uniting subjective processes and objective existences — even when the subjective self is made the object of investigation. To illustrate again : in all of the animal natures there are inwoven, concreted in their organisms, certain instinctive impulsions, some common to all animal natures, and others specific and peculiar to each respective class, as cunning. constructiveness in special forms, as wasp or beaver, ferocity or intensive combativeness ; now thoroughly to apprehend and appreciate these, the self must possess and possesses a corresponding organization of organisms in which is also concreted the same or similar impulsions of organic forces, and so the self thus possesses not only the animalistic impulsions, but a capacity which, while it includes them in this manner, is also susceptible of and combines a more comprehensive and regulative character *of forces*. But the self is, ordinarily, but a spontaneity, until it reacts on its own instinctive spontaneous impulsions, and becomes self-conscious of its motive powers, animalistic and psychical. As the self psychologizes its own action, it gathers and improves its method of psychologizing nature and life. Thus the subject, the objects, and the elements of the self's subjective cognitions, namely, the animalistic impulsions, its actuating, intellective, and affective forces, constitute the intelligence of the subject and the INTELLIGIBILITIES of subject and object; and are thus grasped by the self in the

process of Intusception, that is, by the self going into and
thus comprehending the complement of its own nature.
To be comprehended, all nature must be psychologized,
and until then, nature, or so much as has not been intuscepted — psychologized, will be to the self a dry and
unmeaning mechanism, without intelligible dynamic, plastic, autonomic, instinctive, or autopsic forces ; and will so
remain until the self, in some sense, through its own indicative organisms, functions — psychical functionalizations, grasps the ontology of the inner forces producing
phenomena — facts —*facta*, the things done or made or
in action.   As the method unfolds, it will be seen that
nature and life, with their stabilitations and moving
forces, dynamic, plastic, autonomic, instinctive, automatic,
and autopsic, are but phenomena from a Spirit of the
Universe, and which to be understood are to be psychologized by the self — each self, by culture through its corresponding organisms, from the highest point it can reach
up towards the throne of the Omniscient.   The Spirit of
the Universe can only be psychologized by a correspondent spirit in a congeries of organisms which will bring it
into correlation with nature and life on the one side and
with this spirit on the supersensible side, and thus it can
intuscept that universe and be brought into a realization
— a clear vision — a sense of the action and working-
power of that Spirit in the Cosmos.   God *Knows*; Man
Learns.

3. In all subsisting existences there is the substance —
the substratum of the thing with various qualities which
uniformly characterize it in such combination, that when
we find these qualities in such combination we give the
substratum, or rather in the present state of philosophy
and science, to these qualities combined by their unifying
base, a designate term, as chalk, marble, a rose, a horse,

a man. What the substratum is we cannot perceive or know, as a positive knowledge. The same materials of nature which make the destructive nitric acid, make the life-sustaining air, only they are in different proportions; the same materials which make the grasses enter into the composition of flesh, and the chalk and the marble, are moulded into wheat and bones, and the diamond is but charcoal. Thus are the changing *accidences* of nature seen. Ascending from the plastic forces, changing, constructing, resolving, and reconstructing these elements of nature, and assimilating them for higher uses to the autonomic forces, differentiated into specific germs, life moves forward into the vast families of the vegetal and animal kingdoms, each with their ancillary organisms subordinated to the special typal idea of the species and the individual of each species; but what the differentiated autonomic base of each is, which makes each what it is, we may name in their collective, concrete results, as chalk, rose, horse, man, but do not know their differentiate bases. We do learn and know that there are in them certain well defined and exact correlations, by which, under given circumstances of philosophical contingency, they unite or dissolve, — that there is a composition and resolution of forces, and that they are by such unknown base of germ forces so adjusted and directed, and carried onward, and moulded into form as to produce the intelligible and orderly working of the elements of nature into specific functional organisms, with correspondent and fairly uniform forces and forms. That is Intelligence, in a variety of moving forces, consociated with these forces, and being in itself, as will be seen, a positive force, is infused and incorporated into all the acting forces and stabilitated elements of nature and life. In virtue of these intelligent moving and stabilitated forces, so incor-

porated therein, they act and react on each other, in and
through their correlated adjustments, so as wisely and
well to produce the vast systems of the vegetal and animal kingdoms with the specific and diverse organisms
suited to each and each part of each. This Intelligence,
so concreted and inwoven in all things, when *subjected*
to another Intelligence in this life, possessing a fitting organization for its appreciation, is Intelligibility in objects.
This Intelligibility speaks directly through the symbol
and its agent-forces created by this Intelligence and inwoven therein, to the cognizing — the intellective self —
on this, as it were, the outer and the objective side. The
more elements of Intelligibility which may be incorporated into created bodies or forces, or the more clear
and effulgent the Intelligibilities which are gathered from
the various symbols and their acting forces, to the highest exercise of autopsic Intelligence, the more clear will
be the elements of Intelligibility beaming through them
from the uncreated Forces. Man is the more or less perfect lens converging and transmitting the beams of *light.*
Thus the more certain, as Intelligence in force after
force, symbol after symbol, species and class after species
and class, is intuscepted by the self, will become the
cognizance of the Intelligibility of all created things and
of the primordial Being, until *his* Intelligence will be
seen beaming down — raying out through all our realms
of life, and through the myriad-folded system of the
heaven of heavens. These Intelligible forces will be
seen, foundationally, to resolve into spiritual forces as the
air and the acid are seen to resolve into the same two
identical gases and give an objectifying, creative power,
a redactive, directive intellectivity and an affective, loving Personality which insouls and rules the Cosmos.
" God, a pure Spirit, being the beginning and end of all

things, it is clear that all things, in their beginning and end, must be spiritual. This being the case, material things are phantoms that have no existence, or, if they really exist, they must have their beginning through God and for God, which means that they exist through the Spirit and for the Spirit." — Donoso Cortes, B. II. c. v., *Catholicism and Socialism*.

4. So the more Intelligence in the cognizing agent, the subjective self, with appropriate organisms, and the more perfect the elements of Intelligibility, combining into beauty of form, color, and life in created existences, the more enlarged and perfect will be the knowledge made up of the cognitions by this perceptive, this subjective Self, from a world of such existences. So the more perfect organism given to the human Autopsy, the more perfectly will it intuscept and understand the Intelligibilities as they are concreted in nature and life from the coördinate forces of the creative Being. As the Intelligence in the Self increases, and the Intelligibilities, thus concreted in objects, increase, they are seen to expand through the gradations of dynamic, plastic, autonomic, and instinctive creations, and ascending to the autopsic forces in man; and standing on this summit, this Self will see these simplifying from their manifold differentiations into the trine Coördinations of the objectiv-facient Power, the Intellectivity and the Love of God. Intelligibilities, then, are those qualities inwrought from the Divine Worker by which objects in their actual constitution and in their correlations are and may be understood: Intelligence is that which understands — intuscepts the Intelligible. The Intelligible, it must now be stated, embraces more than the mere unemotive or *uncreative* Intelligence incorporated in nature and life. To make the whole system of nature and life

intelligible, it must be seen that in all the movements of
that nature and life there are always inwoven and moving
forward, in the harmony of their intelligible action, the
coördinates of His objectiv-facient Power, Intellectivity,
and Love *as positive forces.* Thus the highest Intelligence in man or animal, or in the working forces of nature, is but Intelligibility from the coördinate forces of the
Deific Worker; — and thus the messengers of God to
man are those in whom are embodied the highest degrees
of Intelligibility from his Intellectivity, his Love, and his
Actuation — the power to do good, and thus to know and
bear his message — to love, and to infuse love into the
orders to whom they are sent. They each must embrace
an actuating power to do, to act, to objectify forth from
himself the new life of Love and Thought and Actuation
into the lives of others. Thus is such agent the prophet,
the revelator, whether, as in the formation of a new species in the geologic eras, he is a new and distinct creation,
or, as in the order of the successions, he is the product
of secondary causes. In either case as foreseen and provided for he is deific, as such agency. But as he is thus
fully inspired and prepared for his divine work, so are
those to whom he is sent to be correspondingly prepared
to receive; for the message must fail if the messenger
cannot bear it, or those to whom it is sent cannot, from
any cause, receive it. And without Intelligence — mentalized organisms in the tribe or people, the Intelligibilities of the messenger cannot be interpreted — intuscepted.

5. Intellectivity in these its acts of intusception, intelligibilities in the elements of discrimination furnished
in and by the symbols and movements of nature and the
acts of life, in the intelligible forces inwrought into existences in their concrete and adjusted correlations and

interpellences, and as they are inexistent in the coördinations of Being, exhaust the whole scope of human inquiry and cognition, and even that of angelic creatures.

6. CORRELATIONS are those intimate or those possible adjustabilities established, or, at least, which are found existing, between force and force and forces and matter, which give rise to the movements of nature, as they are perceived in organisms, orgasms, functions, crystallizations, &c., as gases produce water, as germs grow into their specific animal or vegetal forms. They are the interchangeabilities of action and reaction. They are the participancy or community which subsists between things of the same kind,' and the adjustable compositions and resolutions of forces between different things. The term is also used for the actual and adjustable intercorrelations of the three spiritual forces in the Self, namely, by which the Self consciously or unconsciously adjusts the willed-power, exercises the Intellectivity, and controls or gives intensification to the emotions. There are, also, correlations of antagonisms; forces repel forces.

7. RELATION is the order in time and position in place (space) of all ideas and ideations as they arise in creation or move into manifestation, and of all existences or symbols or things produced, and the order of their phenomenalizations.

8. CAUSE is potential or actual efficiency. Potential cause is efficiency at rest. Primal cause — *caussa caussans* — is the efficiencies producing existences and *each its own* self-efficient and essential phenomena. Primal cause is the evolution of coördinate forces, or, in its secondary meaning, § 10, a composition and resolution of correlated forces. In the initiate production of existences it may be, on the subsumption of their creation must be, an evolution of forces combining and stabilitat-

ing into *objective forms* and their functional forces. God, as Creator, passes over his forces into an objective — objectified position — into immanence, or else nature is a Pantheism, in the permanent effluence of the divine efficiencies, as Deity making and personally sustaining and operating nature, or else nature is an eternal materialism. On the creative datum, the existence produced may be a composition of forces. When primary elements are thus created by a composition of deific forces, new combinations may take place in virtue of the adjustable correlations inwrought into these primary elements by the constitution assigned to, or subsisting in these elementary substances. Thus new combinations may take place by various combinations and resolutions of forces in or of those elements. In many or most instances of such new or recurring combinations some or a part of the forces incorporated into the elements may be resolved or set at liberty, and be ready, contingently, for other eventual composition. So causes — cause — are evolutions or compositions and resolutions of forces, and always, even in the final effort to reach the foundation of causes, imply plurality, which has been recognized by philosophy, and is the open secret in Trinitarian theology. There must be more than one primal essence of cause. Unity — Spencer's Homogeneity — must begin and end in Unity. That which was eternally a unit cannot, unless in correlation to something other, become other than the same unity — identity — oneness — the same. There must be a force to act and a force, a somewhat, to be acted upon and receive, modify the act — force, and to combine with it. In a universal Identity — simple sameness or oneness, there can be no causation, for everywhere and under all circumstances, if circumstances can yet be predicated, it is the oneness, the same, the old philosophic Identity, the new

Homogeneity. The oneness must remain one; it cannot go over into difference — diversity — multiplication of different identities; it is always the One. Duality, or rather, it will be seen, triplicity, is the fundamental necessity of thought; and this triplicity will be found as underlying and pervading all nature and life. Guyot, in his " Earth and Man," p. 72, says: " All life in its most simple formula may be defined as a simple exchange of relations." Relations, in any sense of this use of the word, are the efficient bearings — the correlations of intercausal differences which one thing has to another, and by which an exchange may be made, and this in virtue of original constitution or secondary adjustabilities concreted and inwoven in the things between which the exchange — the composition and the resolution of forces — takes place. The term, here, " exchange of relations," means an exchange of forces, or a loss and gain of forces, or it means nothing; for he further says: " An exchange supposes at least two elements, two bodies, two individualities, a duality and a difference, an inequality between them in virtue of which the exchange is established." And Sir W. Hamilton, speaking of the causal judgment, says: " The phenomena is this: — when aware of a new appearance, we are unable to conceive that therein has originated any new existence, and are therefore constrained to think that what now appears to us under a new form had previously an existence under others — others conceivable by us or not. *These others (for they are always plural) are called its cause;* and a cause (or, more properly, causes) *we cannot but suppose,* for a cause is simply everything without which the effect would not result." Cited Pro. Log. note C. And Donoso Cortes, *Cath. and Soc.,* B. I. c. iv., under the highest ecclesiastical sanction, says: " The *law* of unity and variety, *that law*

2

by excellence which is both human and divine, without
which nothing can be explained, and which explains all
things, is here shown to us in one of its most surprising
manifestations.   Diversity exists in heaven, since the
Father, the Son, and the Holy Ghost are three persons,
and this diversity is merged without confusion into
unity." And in B. I. c. ii., he says : "The Father is
omnipotence; the Son is wisdom; the Holy Ghost is
love; and the Father and the Son and the Holy Ghost
are infinite love, supreme power, and perfect wisdom.
There unity expanding perpetually begets variety, and
variety in self-condensation is perpetually resolved into
unity." B. I. c. i., he says : "All things are in God in the
profound manner in which *effects are in causes*, consequences in their principles, reflection in light, and forms
in their eternal exemplars.   In him are united the vast-
ness of the sea, the glory of the fields, the harmony of
the spheres, the grandeur of the universe, the splendor
of the stars, and the magnificence of the heavens.   In
him are the measure, weight, and number of all things,
and all things proceed from him with number, weight,
and measure.   All that lives finds in him the laws of
life ; all that vegetates, the laws of vegetation ; all that
moves, the laws of motion ; all that has feeling, the laws
of sensation; all that has understanding, the law of intelligence ; and all that has liberty, the law of freedom.
It may in this sense be affirmed, without falling into
Pantheism, that all things are in God and God in all
things." Yet it will be seen that all this is empty abstraction, unless full force is given to the language in the
beginning, that all these *are as effects in causes*, and that
for *unconscious* nature to be subjected to *laws*, the laws as
forces must be inwoven in its very constitutions.   There
must be actual working efficiencies at every step of this

great evolution, or effects are without causes. Life,
vegetation, motion, feeling, understanding, liberty, are efficient forces, or they are nothing practical, working, effect-
producing; and they are of God or material nature.
This makes it proper here to repeat from the learned
Jesuit, Dr. Weninger, *Prot. and Inf.*, c. iv. § 2 : " You
say, who can believe that in one God there are three
persons? Observing that by the Three Persons we do
not understand three Individuals, but three distinct *relations* subsisting in one nature." [1] The criticism on the
word "relations" as used by Guyot, and on the term "law"
as extolled by Cortes, and I. iv. 16, 17, may be recalled
and referred to; and if Power, Intellectivity (wisdom),
and Love are shown to be discrete forces, or each to be
represented by discrete forces, and these as inwrought
and concreted into all of nature and life, and that these
are a coördinate unity in the Primal Cause, then Science
will have attained significant names for the primal ontologies lying in the coördinate coessentialities of Being.
The differentiations in nature and life cannot be thought
without diversity in the creating, producing identities, nor
can they be thought in their various adjustable correlations, and, most certainly, not in their whole correlations
as a system, without coördinative unity for giving this
system of correlations ; nor are they possible to thought
without seeing the Unity which gives the system and
the Diversity as forces susceptible of differentiations
and capable of being united into system. Thus again
are reached the primal ontologic causes, as Forces.

---

[1] Mr. Herbert Spencer, in his *New Philosophy*, seems to have been
entirely ignorant or regardless, in the assumption of his all-working
Homogeneity, of the doctrine of the profounder philosophic schools,
that Identity cannot produce or evolve Diversity, and of the language of his master, Hamilton, that the conception or notion of cause
is always twofold at least.

9. The first and last question of Philosophy has been an inquiry after CAUSE and IDENTITY — SUBSTANCE. Did Identity produce Cause, or did Cause produce Identity? Around this fundamental inquiry all the systems of ancient philosophizing distinctly revolved. The affirmation was formulated in the Latin phrase, *Ex nihilo nihil fit*: From nothing nothing comes; while in the Greek tongue it was variously phrased, as, To γιγνομενον ἐκ μὴ οντων γινεσθαι αδυνατον: *It was impossible that any real entity should be generated out of nothing*; and again, Ουδεν ουδι γινεσθαι ουδέ φθειρεσθαι των οντων: *No real entity was either generated or destroyed*. Without recalling or reviewing the various subtleties, in ancient and modern times, even to the last essay of Rationalism, which have arisen from this riddle of nature, it may be stated as the result of all philosophizing, that the human mind is incapable of conceiving that where there is nothing to create — no causal efficiency capable of creating, there nothing can be created — there a something will not be produced; and that where there is nothing from which to create, something cannot be produced. Almighty power operating on nothing (such is the solecism of language), nothing will result — no new objective something can be created; the result is nil. But a diversity of eternally subsisting forces — coördinated coessentialities, with a will — power to create, to objectify, with an energizing Intellectivity to create, to objectify wisely and well in intellective forms, and an affectionalizing force, a force emotionalizing to create: — the fact of creation from and by these forces is then not only appreciable, but on the datum that all these forces are spontaneities is a *necessity* to our thinking, just as all other systems or any other is impossible to thought, and on the self-conscious intusception that this Intellectivity

is a *determinative* power, is a moral propriety of thought — a moral necessity of thought. Thus the Self in virtue of its intellective nature, perceiving Intellectivity consociately complexed and interwoven with an executive power of doing, objectifying, making, and with affectionalizing power, and these inwoven in nature and life as correlating and positive forces, reaches the fact of more than one cause of actual causations, and thus reaching them, finds them in their original conditions as moral causes. It is so in its own correlate nature; it ascends and finds it so in the coördinate coessentialities of the Triune Forces. Hints will be given throughout, but the demonstration of the intellective necessity for the positive and actual creation of matters and the diversification of forces will be reserved to the Third Book.

10. Substance, Identity, Cause: these three words, or their equivalents, lie at the foundation of all philosophizing, and are at base but one common expression for the starting-point of all philosophic thought. § 8. Definitions of them — or rather the history of the attempts to define them —writes the various history of all fundamental philosophizing. Cause, it is said, cannot be defined, nor can substance, that is, the substratum — substrata which underlie phenomenalizations, be subjected to any final process of the senses, or to any other methods of perceptive discrimination. We perceive, collect, and colligate the phenomena around some primordial fact, from which we notionalize the ontologic base out of which special phenomena arise. It may be that these bases may be made to stand forth in their naked simplicity, as foundational ontologies for the potential, primary, and secondary causations. The human mind by its very constitution, as finite, is necessarily analytic; but at the same time it must see all

things in *forms*, and it must bind the things which are contained in these forms into the synthesis of their respective wholes, and the whole of the forms into a system of the whole. Now, whatever is known of Substance is simply as a vehicle of cause, and whatever is known of Cause is as a movement — a mover of substance. This will appear more fully hereafter in speaking of mental, moral, or psychical causations; but the facts of nature and the laws of thought require that all motion be seen as a resultant of force, and that one force can only be modified or changed by a correlate or coördinate force. Whatever we know or can think of cause, or of the modifications of cause, therefore, implies and includes a movement of forces. Crookes, the editor of Faraday's "Physical Forces," in the preface, asks : "Which was first, Matter or Force? If we think on this question, we shall find that we are not able to conceive of matter without force, or force without matter. When God created the elements of which the earth is composed, he created certain wondrous forces, which are set free and become evident when matter acts on matter. All these forces with many differences have much in common, and if one is set free it will immediately endeavor to set free its companions. Thus heat will enable us to eliminate light, electricity, magnetism, and chemical action. In this way we find that all the forces in nature tend to form mutually dependent systems, and as the motion of one star affects another, so force in action librates and renders evident forces previously tranquil. We say tranquil, and yet the word is almost without meaning in the Cosmos : where do we find tranquillity? The sea, the seat of animal, vegetable, and mineral changes, is at war with the earth, and the air lends itself to the strife. The globe, the scene of perpetual intestine change, is, as a mass, acting on and acted on by the planets of our system, and the very system itself is changing its place in space under the influence of a known force, springing from an unknown centre." But recognize the simple fact, that the moral forces inwoven into the organization of man must have a physical theatre in which to exercise them, and physical agents through, and by, and on which they shall be exercised, and on the datum of creation, the one will be seen to have been produced for the other; then turn to the minutely correlated forces which give life to and wisely construct the silicious or marble cell of infusoria, eight hundred millions of which are required to fill a cubic inch, and to those stupendously correlated forces which bind planets to their suns into grander systems, and these grander systems into a system swayed by a "known force springing from an unknown centre," the whole filled with differentiated dynamic, cometary, plastic, and autonomic forces, and the conflict and the harmony of these forces colligating the whole into a vast mutuality of correlations, and then contemplate the psychical forces in animals and men, by which these natures are attracted and repelled as among themselves and among their different kinds, and man conscious of repulsion in himself, of his conscious projectility in explosive passions, and of a force which attracts him to others, and of another force which controls, uses, and modifies these forces as well as the physical forces of nature and the psychical forces in other selves, and thus gains possession of the Self's own conscious forces, and it must be seen that moral forces preceded the physical forces to endow them with correlations, to mould them into forms, and arrange their systems and a system of systems. A convolute of the brain, probably not weighing half an ounce, connected by its slight organism with the muscles

of the arm, upon some certain but unknown impulsion of
imperceptible power, adjustedly communicated, will lift
one hundred and fifty pounds' weight, — static force.
Whence the adjustment of the force? certainly not in
the arm; certainly not in the brain, but as certainly in
an adjusting power which used the brain for communication to the arm.   Divide this nerve of motion as near
the brain as practicable, and the power ceases its manifestation; destroy this portion of the brain, and the power
is gone so far.   Forces must have had forces for their
first movement, and subsequent combinations and contrapellences, for their composition and resolution.   The
argument, therefore, which speaks of the creation of forces
as a mere first origination of forces, results in an insolu-
ble contradictory when it concludes that force cannot be
conceived without matter, and that the wondrous forces·
of the universe are, as a new beginning, created.   There
can be *but* three forces, and there must be Three forces.
And these in their Initiate and Final causes are Moral.
God, as creator, and as having a Final Cause in creating,
put forth his moral forces, for until he created there was
no physical cause or force.   This is essential to any ideation of Deity.   His forces are therefore indestructible.
To see this more clearly, effects are seen as movements
taking place or which have taken place : in the former
they will be seen as forces in actual movement; in the
latter they are forces which have moved, and have been
transferred into something else to fulfil other economies ;
while these forces, in rest, are potential forces — causes
to be called into activity.   In this transmutation of forces
is seen their actual indestructibility, and the identity of
their underlying fundamental bases, wherein they are
subjected to, and maintain their laws of differentiation
as well as of original difference.   Their very differentia-

tions must be hinged on the permanence, the indestructibility of the initiate or primal bases, from which the
differentiations were constructed, and run down throughout the system. Centres being given for the action of
forces, there must be a force to throw off from ‑such
centre, a force to retract to such centre, and a force to
combine these two forces, and produce the circle, and in
the circle and these right lines of these centripetal and
centrifugal motions, all figures. It is impossible to conceive one force, mental or physical, as controlled or
directed except by another or other forces. So the
first question in all philosophizing of a fundamental
character — What is Cause ? — must be answered that
it is a movement of Forces. Those who sought a
purely spiritual foundation for the movement of Being
out, objectively, into existences, chose the word Cause,
as a predicate of language without specific and intelligible content of efficient powers — a term merely
covering their assumption of the unknown factor ; those
who claimed the mere material subsistence of the Cosmos, affirmed the eternity of matter with eternal correlations, under a variety of names, exchangeable with
the term Substance under its grosser acceptation : while
the term Identity or its equivalents were common to
both, but more generally adopted by that weaker set
of men who cannot grasp a divine Personality, and
who cannot get rid of the order, system, and harmony
of the determinate movements constantly appearing
amid confusions, and when disorders in the physical and
moral convulsions reach their wildest excesses reappear-
ing and manifesting a controlling power, which is intellective in its ordering adjustments, and which is affective in its love of beauty, holiness, and justice — as He
equates justice. Thus nature, in its geological history,

and in the broad fields of ocean, earth, and air, with their diversified movements and forms of life, presents a vast plane of substances, existing through numberless ages, over which successional forces, according to a distinctive order of succession, exert disposing and redactive powers in building up and garnishing the great temple of the Cosmos. Successionally and as foundational to the other forces, nature was and is subjected to the Dynamic forces, by which were constructed and regulated the star-systems with their planets, comets, and their system of systems: the Plastic forces move upon and through these great masses, thus dynamically framed and sustained, depositing the mines and veins of metal, the mineral crystallizations, and preparing the whole for the later appearance and action of the Autonomic forces, which give to the land, the waters, and the air the vegetal and the animal generations, and unfold into the instinctive and autopsic orders of life. These Differentiations are to be recognized as differentiations of forces, nay, in a final ontology as wholly such; for when the germ-forces of all animal and vegetal life are destroyed, the residuum which is left in the crucible of nature, after the destruction of individuals and species, is composed of some seventy simple substances ; and these again must be resolved, in their final ontology, into forces which have become or were made into stabilitations, or the mind has arrived at the dead wall of materialism. But it must go back of matter, and see that the vast correlations of the seventy simple substances must have been originally constituted for the systematic adjustments of these correlations, giving order, grandeur, beauty, and use to the whole system of things. The mind arrives at the final ontologies of Cause — causes, and finds them coördinate moral forces.

11. SECONDARY CAUSES, *caussæ caussatæ*, are those efficiencies of substances and correlated forces, or of correlated forces, which produce effects, phenomena in nature and life, according to their correlated secondary efficiencies; that is, substances are made with greater or less number of forces of stabilitation to preserve the identities — their specific differentiations, together with forces of adjustable correlations to provide for the change and mutations necessary to the economies of nature and life.

12. STABILITATION is the adjusted equipoise, or an adjustment tending to libration of countervailing forces, according to the correlations inwoven and concreted in them and in their beginning, and for their capacity for composition and resolution, — as the stabilitation of oxygen and hydrogen gas, each, and the adjustable correlations by which they form water, and so throughout the system.

13. PHILOSOPHICAL CONTINGENCIES are those possible combinations and repellences depending mainly on the relations of time and space, § 7, and which, in *one sense*, are not essential to the production of effects by causes. Effects are in the causes, but it depends upon the relations of time and place whether the causes shall unite to produce effects, — as seed is constituted by its internal autonomic forces to grow, yet air, moisture, heat, and light, in proper place and times, must be supplied or may be withheld. Here are all the causative forces in their potentiality; the contingencies are in the facts of supply or not. The facts of supply are essential to the movement of the forces, yet the agency of supply is not in the strict formula of cause and effect, and so the language of Hamilton cited in § 8 is not philosophically correct and precise.

14. CHANCE, in the vulgar estimation, is the causeless happening, not the production of incidents; but Chance as more intelligently understood, is unexpected events, which proceed from the concurrence of several causes, each of which, in its own proper combinations, produces its particular contribution to an effect or result, but which several sets of causes, in and among themselves, are not linked by any natural cohering bond of union, but, on the contrary, are so remotely connected by the relations of time and place, that their coherences or union to produce a result are incalculable to human sagacity or are wholly overlooked, and when the union takes place, by the philosophical contingencies which are at work, the result is unexpected. The result, although philosophically contingent, is yet certain when the whole combination of the agencies are cognized.

15. Another class, called causes, have nominally, or as a motive or end of action, been used for Cause, though possibly not as an efficiency, but in that habitual misuse of language where words are used for things and agencies instead of abstract modes of thought, or, on the other hand, as not giving the full complement of elements and results in the movements of things and agencies. FINAL CAUSE is the reason imputed, or ideated for the divine action, — the why of his action — its abstract wherefore, and in the creation of a thing the abstract, rationalistic *end* for which it was created, and not an Efficiency producing substance or phenomena, and accompanying it to the end, and evolving and reciprocating in its end. Final Cause must have its root of movement as the efficient, the intermediary, and the ultimate Final Cause in the divine love. " Caussa finalis est *id propter quod aliquid fit.* Hæc *influit* in effectum movendo caussam efficientem ad agendum, *amore bonitatis quæ in fine*

est." — Tongeorgi, *M. Ph.*, B. I. c. iii. A. iii. 109, 110.
Final Cause is that, on account of (and for) *which something is made. This flows into the effect, moving* (infecundating) *the Efficient Cause to action for the love of
the goodness which is in the end.* It will become evident
tha͟t Love, *bonitas*, is *in* the Efficient Cause, that *it flows
into and accompanies the effect*, and that it *evolves* in the
end, for the Efficient Love in the beginning. ✒The thing
or faculty is for the love of its use, and its use is for the
love which evolves and reciprocates in some form to the
agent or doer in the end.   Tongeorgi loses the value
and dignity of his own definition of Final Cause in the
exclusive Rationalism to which he confines it in his
further explication and application to details.   Although
he reaches the conclusion that "finem amori *propter*," &c.,
the end is to be loved for *itself*, yet in his rationalism he
loses it as the inflowing of a love which is to reappear
and to reciprocate, in some form, in the end.

16. As in all theologies, and in most of the mental
and moral philosophies, there are various categories of
the Deity set forth, with greater or less perspicacity or
confusion of substantial and fundamental powers, and
which are designated Attributes, of various significance,
yet herein the coördinates of the Divine Being will be
shown to be these coessential, coeternally coördinate forces,
in their essences, several yet united — coördinated in a
common unity of coördination.   They are the sole attributes, the essential hypostases of Deity, namely, his
Power or his Creative Objectiv-faciency, his Omniscient Intellectivity, and his Absolute Love.   Various
modifications of these give occasions for attributions to
God, which are only the partial comprehension of his nature, — as mercy, kindness, benevolence, &c., are but finite
conceptions of his Love ; order, justice, righteousness,

flow from the law — the essence of his Intellective Omniscience; and the human malevolence or wrath ascribed to him is but the working of his mighty Power adjusting the orders of the world. Yet the Power, Wisdom, and Love of God, being always in coördination, these essential attributes always coöperate, and make the unitary whole of his character.

17. Causes are threefold, in the fact that all normal cause is in some sense triplicate, and in the farther fact that the sources of all causes are capable of psychological and rational discriminations into three bases of forces. They are: *a.* Actuocity — objectiv-faciency, chiefly manifested in going out from the self or centre into outward action. It is projectile, explosive, diffusive, centrifugal, throwing off from its own centre of action — objectifying. It is in and of itself, in man, a spontaneity. Pause, and psychologize the Self. There is a discrete class of instincts manifesting in positive force, whose tendency is to outward action from the Self, as the spontaneity of anger and of self-defence, or, when united with the intellective power, of determinate acts of self-defence, or doing or in restraint of action. Yet these complicate with the Love, as in love of self, or love of higher elements of conduct. A close psychology will evolve this power, and this as also *receiving* the other powers, herein eliminated as forces, in all determinate acts, and executing them by being controlled or modified by them. *b.* The Affective force exhibited in attractive appetences; in being drawn to various objects and pursuits, by an affective attraction in the Self. .Do not omit to see that the Attraction is in the Self to these objects and pursuits in life, yet always involving, in a sane and normal state of the Self, the Intellective and the Actuous forces, and that the objects in nature and life are inversely correlated to these appe-

tences. To be so drawn, in the Self, there is a *love*, an appetency, a desire, wish, hope, — an attraction at the base of the affectional force. It is attractive, self-centring, centripetal. In and of itself it, too, is a spontaneity. *c.* Of these two forces, one is projective into outward action ; it is objectiv-facient ; and the other is intractive — attractive in and to the Self. These two forces give all movements backward and forward in right lines, as they give physically all projection, explosion, and all concentration, intraction, attraction ; and psychically are adapted to all spontaneities in the Self, as anger that may spontaneously act outwardly in deed or word, or appetency which may draw to the Self for gratification, regardless of control or direction. But these forces and these directions visible in the physical action of nature, and thus connected psychologically with the instincts or spontaneities of the Self, must be subjected to a form-designing and harmonizing Force, which can control, direct, and redact them into the various forms and uses and economies of life, and of the Cosmos. It is the Directive, Redactive, Intellective Force. This is the self-conscious power which orders, guides, and controls the two spontaneities. Such will be seen to be the correlations of these forces in the Self, where they appear in various diversities of combinations, § 6 ; and as the ascent is made, in psychologizing — intuscepting nature and Being, they will be found to be of, or representative of, the coördinations in God. So man will be seen to be in his spiritual nature, and then he can be seen only as a triunity, and Deity will be seen and can only be known as a Trinity.

18. A *conscious act* of what is termed the Will, as the very phraseology implies, is a complex movement of the *powers* of the Self. The *conscious*ness of the act is of the Intellectivity, and there is in all normal, non-insane

acts an appetency, a desire, wish, a subject-object for the
act. The Act is of the actuous, objectifying force of the
Self. It is of the psychical outgoing of the Self from
its inmost recess — from its own Solidarity, § 34. It is
the dominating force by which the Self, on intensification
from the other two correlate forces, goes out from the Self
in normal executive acts, — objectifying manifestation.
While it is connected with the Intellective force in all
such acts, it is not included in it, but is modified by it,
and it gives to the action its direction and form ; nor is it
included in the Affectional force — the Attractive Love,
the Appetensiveness which attracts, draws to the Self and
gives it the subject-object — the intensification of the appetites, desires, wishes, hopes, pursuits. It acts in conjunction with, and at times in opposition to the Intellective force. The facts of forces appear in the fact that
passions and appetites are sometimes too strong for the
controlling power — force of the Reason, and act with
great violence, and that the Intellective force in turn, in
certain limits, rules them.

Will is therefore a complex act of the Self, and is of
three kinds : *a.* When the Self moves out spontaneously
into action, upon the impulse of some of the projective
instincts, as of sudden anger, sudden self-defence, warding
off a blow, dodging, &c. *b.* When the Self acts attrac-
tively, but spontaneously for the gratification of some
appetite, desire, &c. Here the two Forces are united.
c. The Determinate, the theological Will is the complex
act of the Self in going over into outward action — for the
gratification of some desire, appetite, or affection — upon
selection by the Intellectivity of the appetizing motive,
and upon means, modes, and judgments furnished by it —
and the result delivered over to the actuous, executive
forces of the Self — for they all become such, and so are

implicate in crimes and goodness. For the present it must suffice to say, that the propriety of these distinctions will appear from the confusion of thought always attendant upon the use of this word when it comes to be analyzed,. II. vi., and from the significations of the words translated Will in the New Testament. The Greek word Βουλη, *boulē*, is expressive of an act or state of the mind which was *counselled*, and in the original Greek of the Testament these composite meanings are always present in the use of this word; while θελαμα, *thelama*, as the verb θελω, *thelo*, is expressive of desire, wish *and* act, and the three composing elements are constantly implied.

The necessity of giving an actual content of living forces to the dead words of theology and philosophy is become apparent. Cortes, *id.* B. II. c. iv., says: " In this state of perfect order and admirable connection, all things tend toward God with a determined and irresistible impulsion. Impelled by the law of Love, the angel, a pure spirit, *gravitated* with an ardent and impetuous *desire* toward God as the centre of all spirits. Man, less perfect, but not less loving, was drawn by the same attraction to become associated with the angel in the bosom of God, the centre of angelical and human Gravitation. Even matter, agitated by a secret power of ascension, followed the gravitation of spirits toward the supreme Creator, who sweetly attracts all things to himself." And Dr. Young, Edinb., " The Mystery of Evil and God," c. I. § v., says : " Wisdom necessarily contemplates," and is a causation " to ends, and is determined by their elevation and fitness ; but virtue is its own *end*, and, as virtue, is destroyed by the entertainment, at the moment, of any *end* besides. Moral excellence, of every kind, finds its highest reward in itself alone, and no longer exists, so far as its motives rest on any other basis. And this is preeminently true of the excellence of Love. A generous, benevolent being is one that acts first of all from internal *impulse.* Why does such a being seek the good of others, even sacrifice himself for their sake? No primitive reason can be assigned except the *pure force* of the principle of Love: you cannot account for it on the ground of mere wisdom, mere prudence; if it could be so interpreted, it would then cease to be what it is. He has originally no *end* in view; if he, had, the essential character of his act would be that moment changed (?); a generous, loving nature, *an internal force, to which he freely yields, impels him :* this is the utmost that can be said." Overlooking the want of precision which always accompanies the use of the word "principle," which here, as so frequently, means causation or means nothing, and observing that love without intelligence is only blind and inconsiderate spontaneity, it is beginning to be seen that all psychical and spiritual movements are resultants of actual *causative* forces. If Love is a "pure force," then, as force, it cannot be restrained, controlled, or directed by Wisdom — intellectivity, unless this also is a force. Nor can any phenomena be thought except as resultants of force. Life, nature, the Creative God, moves into phenomenalizations, only in virtue of Forces. And Dr. Weninger says, *id.* c. iv. § 1 : "Whatever God creates, he creates for an end ; else he would act without Wisdom, and would not be God ; therefore man is created for an end ; " but what is this end in the very fact of creation by him, and which induces his Wisdom to create? Wisdom of itself is no *motive* to action, or for adjustment of means to action. There is something more interior still which induces, attracts to the devising of means by the Wisdom to Actuate for it as the *end,* and so the End is an Attraction. The circle begins and ends in Love. God's end for himself is Love; and his end for man is Love; and these must alway be seen as forces in action or for action — attracting, drawing to the end, the object-motive of gratification. § 15. The End reciprocates to the Beginning.

19. Thus it is seen .that the Intellectivity is not included in, but modifies and directs the Actuous — the objectiv - facient force — the objectifying passions and activities. It consociates with the loves in their appetensive, attractive desires, wants, hopes, fears, — selects from them its end of action, gives them forms of expression, and supplies them with means and modes of action in the conduct of life. It is not a spontaneity, as are the other two forces. It can only move in *forms* and *modes*, and upon device or arrangement of means. It is the directive, the redactive force in man and Deity. In man it implies Perception in and through the Senses; it intuitates the Insistent Truth, § 35; and through the Love it attains — ideates, the special, statutory morality assigned by God to man in this special planetary sphere of existence, § 36; and it attains, in like manner, and for the system of the whole, the Divine Ideas from which God fashioned forth the orders and correlations of the Universe. Its province, in man, is to guide, control, purify, and elevate the spontaneities as they arise out of the Actuous and Affectional Natures in man.

20. Love is that remaining element of the psychical and spiritual forces which is embraced in the general terms of attraction, appetency, concupiscence, covetousness, charity, &c. It is broken up and functionalized into orgasms for the animal and human organizations in special and specific forms of affections, desires, wants, appetites, hopes, &c. Refine to the utmost possibility, and when the love of something in all the affectional move-

ments of the self, all wants, desires, appetites, hopes, fears, is taken away, these very elements of the animal and human natures are destroyed. It is the self-appropriating, subjectifying force of the human personality; and its manifestation in life will be as the functionalized orgasms prevail, or the Self reaches to higher loves of Spiritual Life. Man and animals are said to be attracted towards particular objects or pursuits in life; the attraction is in the Self, yet with correlations to the objects of attraction, and yet with adaptations in the objects to gratify the love which seeks them. Their forces correlate.

21. PROJECTILE FORCE, — those forces, however diversified in their formal and local movements, which repel, reject; the centrifugal, projective, explosive, projectile force; and they have but a common root in one force, which throws off from its own centre of centrifugate action. It is seen in the astronomic projectility, but not as a projectility from the sun as a centre, but from a tangential direction, falling into an orbit of which the sun is centre, and in the spontaneous outbursts from the Self. It will elsewhere be shown that *anger*, in a depurated man, as in Paul, the philosophic apostle, is not a malignant emotion, as he clearly saw when he said, " Be angry, but sin not," and that it has no place in any just, or complete ideation of Deity. Malignancy is only a perversion of love; anger in its simple element is only a functional power, proper in men and animals for the preservation of their existence, &c.

22. The ATTRACTIVE FORCE is those forces, however diversified in their formal and local movements, which attract, combine, centralize, centripetate, — the static, the centripetal force. It is seen in the astronomic centripetality, and in the spontaneous appetizing of the

instincts, and appetizings of the human gratifications attracting to their various objects adjusted to them.

23. The REDACTIVE FORCE; the directive, designing, arranging, form-giving, Intellective Force springs from the Intellectivity as its source of origin. It will be seen in the poising and adjusting of the other two forces, in the initiatory planetary and other cosmical creations. In combination with the other two forces, they are bases for all forces and functionalizations. These three forces make the circle; and when seen, the circle cannot be thought as continuously made from himself as centre without the three discrete forces. II. viii. 4–7. These forces, in their simple and complex action, include all forms and motions — all which are possible to thought. To repeat: a carefully conducted self-analysis will show that these forces uniformly spring out of, or at least accompany, in the order in which they have been mentioned, the explosive passions, the affectional emotions, and the operations of the intellectivity. The latter unites with the two former, and gives them form in action and expression, and devises ways and means for both. It is the moulding power in man or in God. It gives *form* to all things. It will be seen as the form-giving force of the Initiate Causalities. When God acts, he must act in some *form*. It is the law of our thought. As the *forms* change, Differentiation is produced. Down in the chemical molecules, lying at the foundations of all known existences, there must be this differentiation. It is actually seen in the correlations and repellences which subsist between things, as it is, intellectually, the necessity of thought that things are different in virtue of underlying discriminate forces or actualities. It is in virtue of the correlations of these primary atomic differentiations that the subsequent superstructures of created

things are built—from the protozoa to the planet. As
the ascending superstructures of nature and life are constituted, Form enters at once into each of the subsidiary
organisms, and into the comprehending Form of each thing,
which includes them, and makes each species what it is.
Each organism, in crystal, vegetable, animal, or man, has
its form, and each its inclusive *form* which thus makes it
what it is. This is the fundamental law of the divine
action, as it is the law of all normal human action; and
thus each *new form* is seen as a specific movement of the
Intellective — the Redactive Power. The acts, the feelings, and the thoughts of men must express themselves
in conscious or unconscious *form*. The conscious struggle of life is to get appropriate Form for the whole evolution of the powers of the Self. Man forms the system
of his life for the gratification of his animalistic, his
human, or his spiritual *longings*. Man forms the system
of his life, — and there is a system of the Universe.

24. FORM must be distinguished from Figure, although
it includes it as the final fact of Redaction. Figure is
simply shape — limitation of extension. Space is limited
by the figure of a thing; the thing is limited by its actual extension. Form is figure and it is quality. It is
differentiation from an original identity, as it is a distinction of one identity from another. The gases have their
respective forms, yet oxygen and hydrogen take a new
form in water, and under other conditions water takes
the *form* of ice. So in all the compositions and resolutions of forces. So in the movements from savage life
to evangelic culture. It is the production of higher
forms for the involution and inweavement of higher
forms of spiritual forces. In the term *formula* it is the
proper or logical expression of thought, as it is also the
verbal definition of a thing or a movement.

25. DYNAMICS are the geometrical forces, by which the respective bodies constituting the varied star-systems, and their subordinate planets and comets, were aggregated according to mass, weight, and distance, and projected and wheeled, and are sustained in their orbital motions. Now let it be seen upon the theory of a *plenum* — space full of diffused cosmical matter, that there was no mass or weight at the orbital distances, and that the sun, the largest aggregation of matter, was yet to be, and to occupy, substantially, a stationary point. That body which should have aggregated the greatest mass and weight would have been that body which, starting with an attractive centre somehow assigned to it, would travel in the largest orbital range through this cosmical matter, thus loosely floating in the *plenum*. But the fact is, the largest mass and weight is at the centre, and where this mass and weight are relatively stationary, and could not be aggregated from orbital motion in such supposed *plenum*. On this theory, at the respective distances of the planets from the sun there was neither mass nor weight, and yet these masses and weights are essential to the very movements of these respective bodies at their respective distances. Without mass and weight in sun and planets they could not revolve in orbital space to gather the cosmical matter, and if they had the mass and weight for their revolution, there is not only no necessity for the theory, but it proves an absurdity. And more; centripetal and centrifugal action, before orbital movement is instituted, must act in right lines, and so there could have been no tangentility in the orbital movements of the planets at their initiate distances from the sun-centre. *a.* The Attractive force, — this is the centripetating, the gravitating, the *vis inertiæ*, the static force. It attracts the stone to the earth as it attracted Newton's

apple and planets to the suns, and suns in vaster systems of orbital movements, until the stupendous systems of an incomprehensible system are orderly " changing their places in space under the influence of a known force springing from an unknown centre." But this is centripetation. *b.* The Centrifugal, — this is the repellant, the radiating, outwardly moving force; and when the force is moving from the centre of origination or location of the force, it is the projectile force. This gives but a straight line in the opposite direction to the attractile force, when speaking of the same centre of forces. They do not give tangential projectility. *c.* The Redactive force, — the form giving force, the force which, moulding the other two forces, at the respective planetary and cometary distances, assigned to them mass and weight, and from them the quantitated volumes with their specially qualitated forces, (planets, moons, rings, zodiacal lights, comets, &c.,) in their great diversity of size and *kinds of matter.* Without these forces in determinate adjustment, from the *embryon* of each cosmical body until they attain their mass and weights at their respective distances, and are wheeled into their orbital planes, there can be no conceivability, either of creation or of the mere formation from cosmical matter, in a *plenum,* of a planetary system. The first force would give simple aggregation, the second would diffuse — repel, and the third is required in itself and to adjust the other two forces. When the planets are wheeled into their orbits, the two former forces may account for the *immanence* of their motions.

" This powerful ever-living agent (Deity), being in all places, is more able to move the bodies within his boundless uniform sensorium, and thereby to form and reform the parts of the universe, than we are, by our will, to move the parts of our own bodies. And yet we are not to consider the world as the body of God, or the several parts thereof as the parts of God. He is a uniform Being, void of organs, members, or parts, and they are his creatures, subordinate to him, and he is no more the soul of them than the soul of man is the soul of the species, carried through the organs of sense into the place of its sensation, where it perceives them by its immediate presence, without the intervention of any third thing. The organs of sense are not for enabling the soul (Spirit) to perceive the species of things in its sensorium, but only for conveying them thither; and God has no need of any such organs, he being everywhere present to the things themselves." — *Newton's Optics*, B. III. 379, 4th ed.

26. PLASTICITIES are those same forces differentiated and adjusted to and upon the dynamic forces, which act on the elements of nature, uniting, attracting, combining, dissolving, separating, and recombining and preparing them for the great economies of nature and life in the earth, the waters, and the atmosphere, for the uses of the crystallizing, and the vegetal and the animal organizations, in their immensely diversified forms. They are the various chemic, mineralogic, and autonomic attractions and repellences and formative forces of these planes or ranges of nature. Yet at base they are seen as the same identical forces, as the centripetal — attractive, the centrifugal — repellent, and the form-giving — the redactive forces.

27. AUTONOMIES are those differentiated forces lying at the base of each individual germ of every vegetal and animal production, and which take up the plastic elements, when all the conditions of their philosophic contingency combine, and then, by these respectively differentiated autonomic forces inwoven in the germs, mould these plastic elements into the different individuals of the various

orders and species of the vegetal and animal kingdoms, and perpetuate species. That is, the autonomic germs are concretions of forces, specifically differentiated, which work intelligentially to the forms of life, vegetal or animal, inwoven in these typal autonomies. What these Autonomies are, cannot be further suggested, than that they must be differentiate correlations of the same underlying forces, and thus endowed with their intelligential and therefore intelligible adjustabilities, by which they take up and carry, as copper is carried in the galvanic crucibles, and deposit and organize the simple elements of matter, and build and form their respective classes of vegetal and animal life in their kingdoms, orders, and species. It cannot but be thought, that the same Initiate Forces which differentiated the tulip and the lily, and the bee and the moth, and the lamb and the wolf, are the identical forces of all nature and life. The occasions of their appearance, in diversified orders, in their respective orders and periods of production, in the very necessity for intellectualizing new forms into concrete existences, indicate specific origins or beginnings in which these differentiate forces were organized for thus acting on the precedent plasticities and the successive assimilations as preparatory to further and higher forms of existences. It is the determination of Intellective Force into concrete Form, yet with the trine forces.

28. Species — autonomies, are the production of life, animal or vegetal, in successions of generations, each in its special differentiated form, by its kind after its kind.

29. These inwoven forces, thus concreted into *forms*, § 25, by which the identity of substances and the perpetuity of species are created and secured, are Differentiation. It gives the organization of autonomies into types. In its higher forms *they* are the wisely working

forces which mould the plasticities of nature into so
many varied forms of use and misuse, and beauty and
ugliness, construct the complex and varied organisms
of the animal kingdom, supply them with their appropriate orgasms, and unfold into them instinctive forces,
and manifest almost conscious powers in the sagacity of
various species, and become, as finite may be compared
to absolute, self-conscious in their likeness to the divine
original, "knowing good and evil." Starting from the
Dynamics, each class of differentiation, as it succeeds in
the orders of creation, is in the nature of newly created
exceptional correlations of forces to that which immediately preceded. The dynamics are the most general
law-forces: the plasticities possess these and something
more; they act in subordination to the dynamics: the
autonomies possess what is common to the dynamics and
plasticities and something more, and cannot act without
them: as the brained creatures possess what those having only a commissural fibre possess and something
more, and in the higher orders a great deal more, until
the simplicity of the forces in their spiritual identities
reveal themselves in man, yet man possesses all the precedent forms of forces in the various parts of his bodily
organization and his psychical orgasms. § 24.

30. INSTINCTS, physiologically, are unconscious, sometimes called automatic, forces of very simple though
differing kinds, as the instincts differ, and in virtue of
which each creature performs those actions to which it is
directly prompted by the impulses arising out of impressions made on its sense-bearing, its afferent nerves, *and*
communicated to the efferent or outgoing nerves to accomplish the end of the instinct, without conscious self-
control or self-direction, and so must be regarded as a
creature performing its part in the economy of nature

from no autopsic direction of its own, but in accordance with the special and intelligential design impressed upon and inwoven in its special organs of instinct, and so in the germ-forces of its autonomy. When an impulsion is communicated to an afferent or *in*going set of nerves, and it is continued around to the efferent or outgoing nerves of motion without conscious check or direction, it is said to be "internuncial" — *i. e.* instinctive.

Philosophically, Instinct is an agent-force, which performs blindly and ignorantly a work of gratification and of intelligence — not of self-conscious knowledge — and in so doing is correlated, in its unreflective knowledge and sense of gratification, to the specific object or objects in nature, which in turn are adjusted to its use and gratification, as carnivorous animals to flesh, herbivorous to grasses, &c. Instincts in their analysis will prove to be the key of the Cosmos — the great temple of Isis.

31. AUTOPSY is the Self, in the more or less clear consciousness of the exercise of its actuous, intellective, and affectional powers, and in the reception of the impulsive instincts from its animalistic and psychical orgasms, and of sensations through the outer sense-organs; and in the possession of a further consciousness of its intuition, § 34, and its ideations, §§ 35, 36, and of its own determinate movements within, to control, direct, or subjugate the impulsions from its animalistic and human orgasms, and to act on nature and in life.

32. As with the Autonomy, instincts are found inwoven and concreted in its animalistic orgasms acting in their appropriate organisms, so for that autopsy is found a brain-organization with organs having their indwelling specific orgasms, and others capable of being charged with special ideational influences, which in their normal and properly cultivated conditions are subject to the con-

trol of the self; yet, when these organisms in some of their
functionalized parts or organs, or in over-tense excitement or undue cultivation, become charged with undue
influences, they react and control the Self, so that its
regulative power is injured or destroyed, as in cases of
ideational monomania, visceral inflammation, mesmerism,
affectional insanities, &c. The very delicacy of organization, which must subsist for the action of the Self
through these organs, might naturally be expected to give
rise to these direct actions and reactions. AUTOMACY
is, therefore, this perverted condition of the animalistic
and psychical orgasms, or some part thereof, and which,
instead of being under the proper control of the Self,
reacts on the Self and controls it by its monomaniac,
fanaticized, diseased, or congenital disturbances. In these
congenital influences it is seen how the sins of the parents are visited on the children to the third and fourth
generation, and how, when they become general in society, a nation, a people, are prepared for their doom
and degradation.

33. As in the Autopsy, in such its environment, there
is a Self standing amid its instinctive impulsions and its
varied soul-organisms, and gathering knowledge from all
these sources and acting on and through them, and they
are thus pouring their diversified facts and influences
from nature and life in upon this Self, each from its respective source, and without order or system, there is a
necessity for the regulative control of this autopsic Self,
to reduce this whirl of instincts, passions, affections, sensations, facts, and knowledge to a systematic method of
life. NORMALATION is, therefore, that regulative and
august power of the Self by which it governs and controls all these various influences, gathers its knowledge
of all kinds, and systematizes all into a method of life.

The bee builds his cell by a most wise instinct, and thus performs his offices as the member of his community; the bird sings his song by a most versatile and charming sense of harmony, some of the species being capable of infinite modulations by an organic imitation which becomes as wise in the native instincts of its songs as the musical automacy of blind Tom, the idiotic negro-boy; but man, having like instincts inwoven in his soul-organisms, builds his houses and sings his songs and discharges his duties as a member of his community, *normalating* them, giving them wider scope, artistic rules, specific intents, and supersensible or spiritual directions, each under a more or less distinct predominance of his conscious powers — of the superintendence of his autopsic self. Development is the prescribed outgrowth from a germ, subject, in all cases to a greater or less extent, to the modifications — varieties, which may be produced by the philosophic contingencies, hinged in the various plastic forces, to act in their times and places, by which it may be effected. Normalation is the tendency to development directed and improved under the inspection and the control of the Self. It is the base of all culture, and the foundation of all moral duties. The progress of the human races is the conjoint production of development and normalation, of spontaneities and sobered second thought. This term, Normalation, excites and keeps alive the conscious conviction of the regulation of the conduct, and of the building up of the actual daily life into a system of life, and, in the higher unfolding of the Self, a life in harmony with the Divine Life. It is the actual and the moral antithesis of the word Development, which in its broadest use has thrown such a withering blight over morals and philosophy, and which has just been revived in the term Evolution, by Herbert Spencer.

34. It is seen what is the Autopsic Self, in its autonomic environment. Through these organisms the heart
beats and the functions of the animalistic life are carried
on; the instincts are enlifed and impel to action and
gratification by their specific and differentiated orgasmic
forces; the higher brain-organism of man gives wider
scope and freer range to functions common to him and
to the animal, as it also gives him the power to write
new forms of thought and action on his ideational organisms, and has given him powers of control and direction
over both his animalistic and his human organizations,
and this so that he may consciously live for the enjoyment of his animalistic gratifications, his human appetencies, or for ascending into the serenity of the moral
order where there is fulness and libration of his spiritual
powers. He may normalate his life within an allowed
circle. In making this ascent, the sharp analysis, at almost every step, takes the distinction of loves in the
animal form and loves in the human appetencies of so
many various kinds, covetousness, ambition, pride, &c.,
and Love, which in a higher ascent, but by inverted action
from this higher point, reacts on the animalistic and
human natures, the love of justice, order, righteousness;
and therefore all along the distinction follows between
the will of the flesh, the will of man, and the will of God.
As the analysis is pressed, (which must now be postponed,) it will be seen further that the fundamental
powers, forces of the Self, and common to the whole of
humanity, are the power of objectifying, doing, making
from the powers in the Self, but always out, over — from
the proper Self, the power of intellectualizing, forming
plans, determining why, when, for this pursuit or that,
and of love, a love capable of indulging in gross appetites, or in the pursuits and the glories of human life,

or of aspiring to a *likeness* with God — to the deobscuration of its own true subjective powers from its animalistic and human desires.   These powers, in multifold diversifications, are the common property of the human races.   It is seen that the diversifications depend mainly on the corporeal and the psychical organizations, and that these are susceptible to culture and to degradation by use and by abuse.   The subsistence in which these fundamental powers inhere — from which they are phenomenalized, being the common property of all the races, must have a common source of origin or supply.   It is herein termed, SOLIDARITY.   The word is not new, and the thought is older than the word.   Solidarity is "Fellowship, or joint interest and mutual responsibility." — *Wor. Dic.*   It signifies "a fellowship in gain and loss, in honor and dishonor, in victory and defeat, a being, so to speak, in the same bottom." — Trench, *Eng. P. P.* 68.   "It is not that all men are merely bound *in solido*; that there is a whole of life, a one life in them all, each individual life being an indissoluble portion of the life of the whole."— Leroux.   Pascal says, "The successive generations of men, continued throughout the ages, should be considered as one and the same man persisting," perduring, "always and continually learning."   Perrault says, "The human race ought to be considered as a single eternal man, so that the life of mankind, like that of the individual, has its infancy, has its manhood, and will have no decline."   The thought lies at the bottom of the theological doctrine expressed in the old formula of Covenants or Federations of God with mankind in Adam and again in Christ.   It implies a universality of spiritual kinhood as the subject and the object of the Covenants.   As such it implies an alterable Solidarity in Adam — the common humanity, by which its nature could have been and was changed, or an alterable autonomy which could affect the doing, thinking, and loving operations and manifestations of the solidarity lying at the base of each Self. An underlying identity of the human race, in its totality, is substantially assumed in this form of expression, and some such subsumption is essential to the moral and philosophical coherence of the theory. That God made a covenant with Adam, the solidaric head of his race, so that in his fall all his posterity fell with him, implies the introduction of actual, effective causes which modified this communal solidarity in those elements of spiritual life, or the autonomies prescribed and assigned to the races; and God made another Covenant with Christ, the second federal head, to rebirth the race, whereby, in virtue of this alterable quality in this common identity of solidarity or of their autonomies, he became a conduit — a means and method, or a method of communication for the same race, whereby all could be elevated who would accept and inwork into their lives *the forces of alterability*, and which would so affect, directly or through the depuration of their autonomies, this solidaric element, this common bond of identity underlying the races, as to bring the individual Self into harmony with God through this federal head. In the Catholic church, "It is the living and organic Unity of humanity," mankind, "in virtue of which each shares a responsibility in common with others." — Cortes, B. iii. c. 2. These views and statements are hypotheses or descriptions and accumulations of phenomena gathered together, and reduced to some theory of the life of mankind as observable in its natural history. They are collected from writers of the most extreme divergence of philosophical and religious tendencies, and indicate that constant desire for unity which pervades all earnest

4

and inquiring thought. The concurrence of the philosophical and religious formulas will appear in these respects: that in mankind there is a communal solidaric element located in the diversities and depravities of their respectively organized autonomies; that in virtue of the very essences of the solidarity, which appear wherever we touch them in life, or ascend and think them in God, there are bonds of sympathy and union; that in the differences yet kinhood of the autonomies there are bonds of mutual affinities, dependencies, and correlations, yea, and in virtue of the disproportions inwrought in their organisms and operating in the economies of life, repellent antagonisms; that the degradation of each individual, thus manifesting and intensifying these antagonisms, tends tò degrade others and produce corresponding manifestations, — thieves, drunkards, murderers, prostitutes, &c., herd together, and the exercise of one passion or malignancy excites the respondent natures in others; that the depuration and sanctification of each individual tends to communicate depuration and sanctification to others, in their concordant association; and as each ascends he suppresses or loses the respondent natures of the lower classes. Man does not instruct his passions and affections so as to give them their original direction of action. These are inborn, yet he can intensify or control them by his more or less determinate action. But his ideates — his opinions, domestic, social, political, and religious, are the results of his position in the time and place wherein he appears in the movement of the prolepsis. These, the passional and the affectional orgasms, and the ideational capacity, and their appearance in the family, tribe, and nation, give the distinctions of individual, family, tribal, and national characteristics. The underlying elements are all the same, yet there is ever a movement and change. Thus all mankind are as it were afloat "on one bottom." Solidarity is that communal immanence of underlying identities, complexed in a unit of consciousness, out of which are evolved the triplicate phenomena of consciously doing, consciously intellectualizing, and consciously loving, (in the highest form of love,) and all these are imperfect and truncated, except in the unity of their highest fundamental excellence. It is the immortal Identities of the human spirit. There is no conception — notion, of identity, substance, or cause below these, more fundamental than these; and though they may differ in manifestations, in individuals and races, in degree, they do not differ in kind, and the fact of their Identities are essential to any system of morals, any progress of the races, and any continued responsibility, now or hereafter, for human conduct. Light, composed of three simple elements, in passing through various mediums is obscured, separated, or broken into limitless shades of color; and its uses in the economies of nature are infinite. Around this solidarity are builded the autonomies of individuals, and of the races, to be effected by the moral causes which degrade or elevate the psychical organisms, or to be broken, injured, or destroyed by the causes dependent on the philosophical contingencies of nature and life, and the moral action of individuals. Herein the divine system, as it works in the ages and the normalation of individuals and tribes, as they crawl and creep or walk erect through the divine plan, are seen; and man blindly gropes in the darkness of his depravities, or more or less openly aspires to the loftiest heights. "The dignity of the Image is seen in the first state of man, but the perfection of the likeness is attained in the Consummation."

35. As this Autopsic Self starts from its envelopment

in its special Autonomy and the obscurity or light of its
relations in time and place, and advances in its cognitions,
and attains the self-analytic consciousness, it reaches a
realm of Truth, called by various names in different
systems of philosophy, as the Immutable Truth, the Pure,
Absolute, Impersonal, Perfect, Universal, or Eternal
Reason, or the Eternal Principles of Nature. In special
systems it is called Mathematical or Geometrical Truth.
It is herein called INSISTENT TRUTH, when spoken of in
general terms, as being those truths which we must
perceive or Intuitate as eternally insistent, whether man
was made to intuitate them and act upon them or not,
or matter was formed in which to embody or symbolize
them. Two and two are four, and in every right-angled
triangle the square which subtends the right angle is
equal to the squares which contain the right angle; and
all such propositions are true, and must be intuitated as
true; though matter, man, or Deity did not subsist. Without stopping here to point out the want of rigid analysis
in the following language of Sir Wm. Hamilton, and his
failure to discriminate these objects of the Intuitional
power, his Noetic Faculty, from the Ideas which are at
the end of the objective processes of his Dianoetic Faculty — properly the Ideative function of the self, — in other
words, the want of discrimination between geometrical
truth and those Divine Ideas on which God patterned the
forms of all existences and their correlations to each
other, — the authority is used to show the general, if not
the universal recognition of the Insistent Truth, under
a multitude of names and terms, but which do not discriminate the one from the other. He says: "Were it
allowed in metaphysical philosophy, as in physical, to
discriminate scientific differences by scientific terms, (! ?)
I would employ the term Noetic, as derived from νους, to

express all those cognitions that originate in the mind itself; Dianoetic, to denote the operations of the Discursive, Elaborative, or Comparative Faculty. So much for the nomenclature of the Faculty itself. On the other hand, the cognitions themselves, of which it is the source, have obtained various appellations. They have been denominated, κοιναι προλεψεις, κοιναι εννοιαι, φυσικαι εννοιαι, προται εννοιαι, προτα νοηματα ; naturæ judiciæ, judicia communibus hominum sensibus infixa, notiones, or notitiæ connatæ, or innatæ, semina scientiæ, semina omnium cognitionum, semina æternitatis, zophyra (living sparks), præcognita necessaria, anticipationes ; first principles, common anticipations, principles of common sense, self-evident or intuitive truths, primitive notions, native notions, innate cognitions, natural knowledges (cognitions), fundamental reasons, metaphysical or transcendental truths, ultimate or elemental laws of thought, primary or fundamental laws of human belief, or primary laws of human reason, pure or transcendental or *a priori* cognitions, categories of thought, natural beliefs, rational instincts, &c. &c." — *Met. Lec.* xxxviii. It is needless to translate specifically any of the terms used ; those in the English language are substantially the same, and show the object and subject — the thing sought with the thinker seeking, and they imply immediate cognition — intuition, as also his processes of cognizing ; but these and the whole language show that there is a realm of Insistent Truth which the mind Intuitates — sees by its native Intellective light. *They are simply of the Intellectivity.*

36. There is another class of Truth, not simply of the Intellectivity, nor will the Intellectivity *alone* ever give it. It presupposes other elements than can be found in the Intellectivity : it requires, in thought, a moral nature

in God, and necessitates a moral nature in man. And
morality is a *love* of order, justice, righteousness, for the
sake of loving natures. Much, in many books, has been
said of Immutable Morality. The immutable morality
is *the law and life* of the deific Ruler. What his morality is no one can say, other than that it is in some inscrutable way the ultimate coördination of his Power,
his Wisdom, and his Love. The attempt to form a system of absolute, Immutable Morality for the Divine Being
will result in inexorable contradictories of the profound-
est philosophical and moral significance; why did he
permit evil, vice, and sin? why did he make man to suffer
through all these dark and gloomy ages, and not unfrequently the best to suffer the most, as if truly the blood
of martyrs was the seed of growing holiness in the races?
Therefore let us be meek, and submissive to the *appointed and statutory morality* he has assigned for us in
this planetary theatre of our existence. As the system
unfolds, it will be seen that if this earth had been nearer
to or farther from the sun than it is, the organisms of
the animate natures inhabiting it would have had to
have been different. Accordingly the laws of their
physical condition and action would have been different.
So if the human organization had materially changed
to conform to a different sphere of existence, and
dropped some of its present organisms, or was differently correlated to others, the laws of the moral action must correspondingly have changed. As Christ
said, "*there* there is no marriage nor giving in marriage," then there none of the moral laws relating to
marriage have any application. And so through the
whole moral law. Where the Spirit is immortal, there
is no necessity for property in its various uses, and the
many moral restrictions and injunctions relating to prop-

erty would have no application. Where the Spirit is immortal, there can be no murder. There there would be but one element of the Decalogue left — Thou shalt love the Lord, thy God; but when this love, by the gratifications of our animalistic and human natures, is turned away from God in the conscious and uniform violations of the other portions of that law, and their loves are inwoven by the indulgences of life in the roots of our solidarity, the resurrection can only be to profounder perversions and malignities. As there is a proleptic order and movement for humanity through the ages, let us meekly and wisely submit to his appointed statutory PROLEPTIC MORALITY.

37. There is another class of mental phenomena attained by the Self, called, also, "reason," "ideas," "transcendentalism," and which may be properly termed DIVINE IDEAS. These are reached by a process herein termed Ideation, the intusception by the Self of the divine *forms* from which the movements into creative actualizations were objectified by Deity. To realize, to ideate these infinite forms with their functions and correlations, as they appeared in their relative successions in time and place, requires the concurrent use of the triplicate powers of the Self. No one or two of them will give the life-content. The intellectivity *thus* goes and sees that the Forms of things are selected from the divine omniscience, and objectified into nature, as it thus grasps the love which enlifes them and the powers which actuate them in their coördinate movements of creating. This is reaching the Transcendental, the Divine Ideas.

38. The first step in this process is ANALYSIS. Analysis, in practical life, is to take that which is in its whole, its entirety — its synthesis, its Form, and dirempt it, separate it apart, piece by piece, element by element,

component part by parts of any machinery, organism, or compound substance, and observe the materials, elements, forms, functions, forces evolved, and uses, for the purpose of understanding each in its part or discreteness, and in their relations and correlations to the whole, and thus understanding the whole or so much of the whole as is proposed for examination.

Analysis, in the personal life, is the observation of the content of a complex state of the mind or of a collective ideation, or of a· psychical process, or of a spontaneity which has been recalled, or of an instinct which has appetized and gained its gratification, and the separation of their phenomena in their differences and diversities, and their modes of action as parts of a complex whole, and as they appear and act as simple forces. In all cases it is divulsion; it is throwing or putting, and in a sense projecting, the parts of a thing from its own centre of cohesion or combination. It is anatomizing. But the Powers in God, or in the Self, do not act separately, except in the Self in manias, dreams, or reveries ; and here they are present in some of their varied forms; and neither mind, nature, nor life can be understood, except as they are psychologized, as the Self reënlifes them and pours through their organisms from its own inner Self the movement-forces of their construction, and sees their repulsions, attractions, and their redactive processes. It must see the composition and the resolution of forces and their forms.

39. Analysis is but ruin and desolation. SYNTHESIS reconstructs — draws, brings, attracts together. Synthesis. in man is thus the bringing together the parts and cognizing the mental rule, the fore-plan of the *modus operandi*, of the method of constructing, forming, or making, and working in any given finite constructure, or the

transcendental idea which ruled in the formation or
transformation of any force or the creation of any element or organism. While there may be a chance dis-
covery, § 14, of such rule or transcendental idea without
previous analysis, it is only *guess*, and can be verified
but by subsequent analysis. There is not and there
cannot be such a thing as intuitive synthesis by man, —
a law or idea without a knowledge of the facts on
which to base it, — by and through which to intuscept it.
Power can only be intuscepted through our own Actuation; Intelligence, intelligibility, only be known through
our Intellectivity, and affectional natures through our
own affections. One Self may require a less number of
facts of the respective kind, or a less time for considering the facts, than another; but the life, the moving
forces of the facts, must be attained in the same manner.
As the analysis takes apart, piece by piece, &c., the synthetic cognition unites, attracts, draws, brings together in
their order and coherence or combination of correlated
parts, and conjoining and actuating forces.

Pure Synthesis belongs alone to Deity in his Omniscience. The ideal synthesis of the Cosmos is the tran-
scendal picture of that Cosmos as it lay in the Divine
Mind, before the creative act — before the ongoing of the
forces moved into an objective position and state, and
stood, as it were, in stabilitated immanence and manifestation over from the Divine Self.

40. In these processes, analysis separates, and the Self
cognizes the parts and the forces in their separate forms
and qualities. Synthesis is the Self intuscepting the
reconstruction part by parts, and, where parts are already
known, by parts in greater aggregations of knowledge,
and the Self pours through the parts and the whole the
forces which move them, until the unity of the whole in

its elements, organisms, and forces lies clearly in the Consciousness. Synthesis is the drawing together in reconstruction and enlifement of that which the Analysis has anatomized. And when the elements, organisms, and forces are thus brought together and re-enlifed, and the appropriate *form* is bestowed, it is REDACTION. Thus the Self starts with form, finds *forms* at every step of its processes, and returns to the final *form*. So man can only ascend in *forms*, or to find *forms*. § 24. As the *parts* of a whole have each its *form*, so must the whole have its characterizing Form. Of those things which can only be seen in mental vision, the formula can only be one of language, and this is given in common and philosophic modes of expression. In the actual fact of creating, creation begins with analysis in the preparation of elements in their simple quantitative particles and qualitative forces, and ascends by compositions and resolutions of forces to the redactive forms of the Divine Ideas. §§ 23, 24. Deity descends from his omniscient synthesis to this minute analysis in created nature. Man begins in analysis, and ascends towards the divine synthesis.

41. TRANSCENDENTALISM, then, is the ideation in whole or in part of the deific system as it preceded creation, and before it was materialized and actualized, and actualizing in the universe. It is the intusception, as far as it is given us, of the foundational forces in their laws, capacities, and forms as assigned to the Cosmos before they were and are actualized in positive, concrete creations. This so, then the path of investigation — the *way and the life* for obtaining a proper apprehension, a re-enlifement in our own selves of the system of the universe — is by intuscepting that universe in all its wisely working and affectional powers through the triplicate

foundational forces springing out of the common solidarity of the race, and which testify to each one, more or less, in his accorded time and place. These alone can give us causes in their potentialities and efficiencies for furnishing forth a creation, where the dynamical forces present those possessed by the Self, where the plasticities inwoven by the vegetal and animal autonomies work blindly but wisely intelligential to their forms of beauty and use, where the instincts of the animal creation incorporate a power — a force for acting, a blind intelligence to fulfil their appropriate offices, and evolve a love with diversified means and sources of gratification, and where the Autopsy, which makes the transcendentalism, exhibits, in its spontaneities and ideational reveries, evidences of identification with these forces, and in which the self-analysis discovers their conscious inherence. It is now averred — but the truth will· constantly break forth — that the Primal Causes are triplicate, and originate in the coördinations of a Triplicate Unity, common alike to God and to man, but only likened and imaged in man to those in God. The man of mere spontaneities cannot make the intusception. He lives only in his passions and impulsions, and these he has not analyzed; but he of consciously normalated life, standing, as man may, in some sort, above his passions, affections, and even the cunning and sinful working of his intellectivity, can catch· the spiritual insight, and see the Cosmos in its beauty of forms and grandeur of movements. This system, beginning in God and moving to his ultimate purpose, his Final Cause, is the Divine Prolepsis. § 15.

42. This PROLEPSIS is the actual working of the intellective fore-plan selected from and in the omniscience of the Deity. It is the movement and actualization of the Cosmos as preordained before the foundations of the

worlds. To speak of this planet, it is the movement of
the forces concreted and inwoven into the processes of
nature and life, and on and by which they tend to and
produce the successive conditions of things in each separate department of creation, and the Final Cause in the
syntactic whole of nature and life. The Final Cause in
this Prolepsis is the state or condition of existences,
which they are to bear to their originator in some end
or prolonged system, which is the intentive object of the
Creative Mind. As this end or aim can only be attained
by intermediary forms, and differentiated forces, in a vast
compages of converging correlations, and which forms
and forces, in virtue of the intermediary, remote, and
final end to be attained, must be formed and differentiated to work wisely and well in their appointed spheres
of correlations and repellences to produce such ends or
aims, so the forces are so inwoven and wisely correlated
to converge to and produce such intermediary, remote,
and final ends. Thus to work, to actuate, these intelligible causations must have been inwoven, from their
beginnings, into the practical efficiencies of each existing thing, and each part of the system, and the whole of
the system. They imbue and pervade the entire compages of the Cosmos, and give them action, and limits of
action, to produce and secure the arrangements of the
prolepsis. This prolepsis is visible in the linked series
of cause and effect in the physical world — in the indissoluble connections so constant in the economies of nature.
Herein it is so palpably visible, so invariable, that many
minds deny the necessity of a creator, and affirm its
eternal insistency and ongoing. But the philosophical —
the preordering prolepsis is visible in the order and evolution of higher and clearer results in the geologic successions; and the moral prolepsis is unfolding its argu-

ment in the conscious use and control by man of physical causes and effects, in the higher conditions of intelligence and moral struggles of the races of men, in the efforts to adapt governments to the wants and improvement of the individual, and in a reaching after perfection, which will make governments useless as but instrumentalities of a lower condition of life. The Prolepsis, then, is the means and the given end of the creation in and for the Creative Power, Intellective Omniscience and ever-working Love of God; or, in the language of a severe theology, " We shall rejoice at seeing the Wisdom, Power, and Goodness (love) of God are infinite; these are the three attributes [§ 16] principally displayed in the creation; the universe is their work, and may be considered a short essay on their perfections, or *a piece of mechanism in which they are wonderfully displayed.*" — St. Francis de Sales, *Love of God*, B. IX. c. i. If not now, as the argument unfolds, the serene and eternal beauty will come forth of that passage quoted in § 8, " The Father is Omnipotence; the Son is Wisdom; the Holy Ghost is Love; and the Father and the Son and the Holy Ghost are infinite love, supreme power, and perfect wisdom. There, unity expanding perpetually begets variety; and variety in self-condensation " (in its central attraction) " is perpetually resolved into unity."

43. If, then, these three are the only forces, — the actuous, objectifying force to be distinctly psychologized, when the Self goes through its psychic organization, and acts on muscular integuments, in determinate action, to do, to execute, to perform, and thus evolves physical force or its exact equivalent in putting physical force into action or overcoming it, and in instinctive acts of self-defence, and in spontaneities exhibiting deeds of anger and wrath, §§ 17, 18; — the affectional force, when

the Self loves, desires to attain, hopes to accomplish, to attract, enfolds, holds, maintains, conserves, although the love combines more or less with its other correlate forces, to obtain its various gratifications in the various actions and economies of life; — the intellective force, which selects this motive of gratification, and rejects that which devises ways and means, which gives and *guides* into forms, modes of actions, moulds symbols, and gives specific manifestations: — there being only these three primal essential forces from which to furnish forth creation, and these being accessible to us in the Consciousness of the Self, when the Self can, with less or greater clearness, follow the movements of these forces, and feel and legitimate that it is by these or consubstantial forces that a planet is aggregated without mass and weight in its initiate construction, until its mass and weight are properly adjusted in its exact relations of time and place to its sun and the other planets and star-systems, and then projected on its tangential course, and retracted into its orbit, and performs its located motions; by which all bodies are drawn to each other, and to the centres of their planets, yet with a constant repulsion interposed; by which the plasticities attract and repel, and form and mould the elements for organizations according to their directive correlations; by which the autonomies, in the vegetal and animal Kingdoms, in their manifold differentiations, mould those plastic elements into forms of horror, or beauty, or use; by which the instincts do their deeds of active power, yet of blind intelligence and love in gratifications; and then sees, crowning the summit of the movement, the Autopsy in which man exhibits these forces in his own actions, affections, and thinking, the Self can intuscept the universe, and declare his image as after the Image of him that made him and all things;

and as he rises above the narrow limits of his organisms, and is depurated from the impulsions of his animalistic gratifications, and his human desires and malignancies, and becomes conscious of " the gift of God that is within him — of Power and of Love and of a Sound Mind," he will be not only the Image but the living and incorporate Likeness of him who made him after his own counsels. § 25. How the solidaric Self is connected with the organic forces, and moves, as it were, along their courses in functionalized organisms with their impregnate orgasms, can only be experienced in conscious watchfulness of the movements of the Self in the self, and in life, and into its correlations with nature; but its sharpest and only full introspection can be given in that terrible self-analysis in which the new birth into the spiritual life throws its broad strong light into the gulf which separates spiritual purity from the animalistic and the human depravities, and the Self unfolds its spiritual Powers for their control and subjugation. The Method of this insight, and the insight itself, to some extent, will be unfolded in the normalation of the facts of consciousness common to cultivated and pious minds, and from the conceded facts of the philosophies. Thus passing back introvertedly and analytically over the movements of nature, and the articulations of the Prolepsis, the Self will reach the transcendental synthesis as it moved out of the omniscient design into the forms and forces, fulfilling and manifesting the system of the universe.

BOOK FIRST.

THE GROUND PLAN.

—◆—

CHAPTER SECOND.

THE ELEMENTS, FORCES, AND MODES OF CONSTRUCTION.

1. ALL objects and subjects of knowledge and ideation fall into two great divisions: Being; Existences.

2. Being, in and of itself, is not an object of knowledge. It is an object of Intusception, and it cannot be
reached by a formal process of reasoning — of notionalizing as of an ontological subsistence. It is to be intuscepted and sought for as it is revealed in phenomena,
and these phenomena are to be found in all elements
which are subjective in or objective to the cognizing
agent, — in the infinitesimal protozoa, the worm at the
feet, man in the mystery of his great and depraved multitudes of the successive generations; in the star-systems,
in the nebulæ at countless distances, in the gleam and
glory of the more highly gifted thoughts which flash and
fade; in the more slow and solid structures of architec-
tural intellections, which mass up in the soul through its
years of labor and diligence; in the lowest instinct, and
in the highest consummation of Love and Thought and
Action embodied in a personality, as a Purifier of the

nations ; and, gathering these in their *concrete forms,*
sparkling and burning with their inwoven intelligibilities, ascend, transcendentally, and see them as necessarily foreplanned and prearranged, — and we will stand
in the august Presence.

3. BEING, incapable of disjunctive analysis, subsists
in a coördination of trinal coessentialities, namely, Power
to create objectively, to sustain and to destroy what is
created ; they are equivalents : Intellectivity, the *forma
formans,* to correlate all things, from the primary elements on through to the last act of time, to wisely organize and adjust the positive correlations in matter and
forces from which cause and effect flow, to counterpoise
the philosophic contingencies, filling a space or conditions
in the economies of nature and life, but less important
than the actual causations, I. i. 13, and to supply, in
these and other manifold forms, Intelligibility to nature
and life, and Intelligence to the workmen and the worshippers in the temple ; and a causative love *from* which
and *to* which all shall conform. I. i. 15.

4. Physical nature with its perduring stabilitations
and its flowing, changing phenomenal manifestations ;
psychical nature with its functional activities ; spiritual
nature with its presidency in and over the animalistic
and psychical orgasms and organisms, (for there are
orgasms which impel to action, and there are organisms
on which the Self inscribes its own characterizing notations,) all interlinked in constant and yet ever-changing
correlations, and nothing steadfast in this mighty sea of
change ; — the whole attracts our wonder, and demands
the solemn exercise of all our power, our wisdom, and
our love, that we may know, and love, and do.

5. Existences are of Body, Soul, and Spirit, II. ii.
Σομα, Ψυχη, Πνευμα of St. Paul. I. i. 1.

5

6. Body, matter, is posited as the base of what are commonly called inorganic substances, and as the menstruums or vehicles of the forces which make, or make and move, organic existences of the physical and the psychical natures. Not repudiating the theory of Boscovich, revived by Faraday, but maintaining that all existences are forces in stabilitation or play of forces, yet, as the processes herein instituted lead to the ideation of Being, as coördinated in these trine coessentialities, and as the maker of differentiations in the primary elements of matter and of autonomies in vegetal and animal life, in organisms and functional orgasms as subsidiary to his Forms, I. i. 25, 29, with independent and interchanging forces, body,—bodies must be accepted in a scientific ontology as the menstruums or instrumentalities for the play of the specific functionalities and the mediums of the interchanging forces. The theory of Boscovich and Faraday as a base of *positive* science must be set aside, yet as a base of philosophy for *created* existence it may be, is undoubtedly true, if God is spirit and matter is not eternal. I. i. 8–10.

7. Soul is the organisms in which are inwrought various functionalizations, independent in their respective powers, yet correlated to muscles and viscera in the corporeal frame, and organically arranged for repercussing from one to another, and in a normal condition subordinated to the control of the spiritual autopsy. I. i. 1–3, 17–23. It is mainly in the exercise and control of these, in the vicissitudes and contingencies of life, that the responsibilities of the Spirit are hinged and arise. The fox is cunning, because he cannot help it; it is inwoven in his brain-organism and acts instinctively, as monomania in man is automatic; man is cunning, but shall he use his cunning for fraud, or to make his destructive

orgasms more effective, or shall he use it to prevent,
control, and counteract these, or lead others to prudential or wiser conduct? The Spirit rules or is ruled.

8. The evolution and elucidation of the Triplicate
forces, as they arise out of the autopsic solidarity of the
Self, will occur in the normalation of the true elements
and correlations of life, and will irradiate their light in
every chamber of the great temple.

9. Inorganic bodies are those of greatest immanence,
stabilitation, and which are not under the actual influence
of or which have not been concreted into some functional
form by plastic or autonomic life. Yet, all matter is
organized; it is a congeries of elements held together by
their concreting and correlating forces.

10. Organic matter is body functionalized into formal
organism. The greater the variety of functions, the more
complicate the organisms.

11. Organized bodies, as organization is commonly
denominated, are divided into Botany and Zoölogy, although Mineralogy and Metallurgy are other orders of
organic and constantly organizing forces, — the organic
forms in both, in many instances, being distinct. The
vegetal orders under the control of their variously differentiated types, by virtue of their autonomic forces, build
up their own organisms, and complete the specific end of
their lives in the production of seed or germs for the
continuance of their species. They are correlated to
other uses in nature, and in turn contribute to functional
powers in other departments. So in the animal world ;
but in the ascent of the gradations, animals are correlated to still higher uses, in more intellective functions,
superimposed on their lower instincts, as in the dog, the
cow, the horse. These intellective functions must have
their organisms of power and use, so as to apply their

semi-intellective actuation and love to the master's use.
And soul-organism is given them of its kind. As instinct
produces respondent action of the *motor* nerves from the
influence communicated by the *afferent* nerves, and continued without consciously determinate break into the
motor nerves, so the souls of beasts, in their semi-con-
scious obedience and love to man, are in a certain way
automatic, and act from these instinctive and repercussive
impulsions — from the nature inwoven in their respective
organisms and not from the presidency of an autopsic
spirit normalating their conduct. I. i. 3. In man the soul-
organism is more perfect and distinct, and the influences,
impulsions carried by the afferent to the motor nerves,
are capable of being consciously and self-determinately
broken up, while it may and frequently does act from
instinctive and automatic impulsions, as in the more bru-
talized or impulsive members of society and in monomanias and constitutional idiosyncrasies ; yet, in certain
limits, personal to each self, the autopsic Self controls
this automacism by the human culture of life, or by an
appropriation and inweaving of the spiritual forces into
the organic action of life. " Who knoweth the spirit of
man that goeth upward, and the spirit(-soul) of the beast
that goeth downward to the earth ? " — Ec. iii. 21. The
language of that day could not convey the discrimination.

12. Animal life is arranged into four orders, if the
infinitesimal protozoa do not present any other. At the
base of animal life are found these infinitesimals of life,
composed of infinitesimal atoms still more infinitesimal.
The successions follow in their diversified successions,
— the Radiates, the Mollusks, the Articulates, and the
Vertebrates, — and each order in ascending forms of
perfectibility ; and man is classed in kindred by dis-

criminating elements of relationship, yet of irreconcilable difference, with fish and reptile and bird and
beast. This is so articulately indicated, that man, standing at the summit of the vast creation, can in the processes of his intusceptions, passing down through himself,
see the long lines of existences, and catch the moving
forces which actuate the individuals and the orders beneath and around him.   He intuscepts, from himself,
their varied forces to actuation. Animals of the same
species understand each other; nor animals nor men
understand the other, except in and through those portions of their nature which they possess in common. So
man, passing down through the long geologic eras, realizes
the deific activity in all forms of existence, working
countless ages from various intercalated beginnings, indicating and converging to a beginning, interpolating differentiated organic forms in their times and places, embodying more varied organic forces, and crossing the chasms
and breaks of the geologic dislocations, ever with the
higher forms and more varied functions, and tending
always, in a clearer prolepsis, to man — to man, the conscious symbolization and predetermined autochthon, embodying in his ultimate and perfected consciousness the
triplicate powers by which were created and from which
were furnished forth the whole. At each geologic break
an interpolation of the ascending gradations, in new
forms and functions, takes place ; and these more intelligibly characterize the various cycles of creations, increasing the distinctness of the agencies at work, until in
the intelligibilities incorporated in man a power of doing,
making, creating from his own forms, actuating for his
own loves, is set free, — an intellectivity for the first time
in this series of creations grasps and follows the lines of
thought pervading the manifold complexity, and Love,

heretofore with his face to the earth, and dwelling among the fierce gratifications of the monster-periods, now turns his thoughts to heaven, and adores in grateful veneration. Love, the redeemer, the purifier, the reconciler, — last born into the open order of time, its triumph will be the subordination of all loving, conscious personalities. But geology repeats itself in history; man has had his monster-periods; and the corresponding types of the tiger and the monkey murder and mow and chatter in the orders of men.

13. This triplicate Self is the only creature which can interpret nature, and precisely as he has an enlarged consciousness of actuous, practicalizing force, of intellectivity, and of love, his capacity to comprehend nature is enlarged. Without this practicalizing force he cannot comprehend the operations of Actuous power; without love he cannot make any intusceptions of love in any of its broken and functionalized gratifications, nor ascend and know it in God; and without intellectivity he cannot comprehend aught. It is, therefore, only as the three are brought into harmony and coördination by an auspicious organization, or the moral normalation of life, that the intusceptions of that which is above us can be complete, while of that all around, they will be imperfect. § 2. Without organism to bring the self into collisions and harmonies with all nature, there can be no communications to or from nature, so the organisms of man give him his consciousness in nature, and place him in the web of complexities, wherein are constantly evolved the trials and responsibilities of life.

14. The solidaric Self, in the personality of man, is the conscious autochthon making the investigation. Under the activities of the Self is the philosophical and empirical fact of Consciousness. The simple and yet

complicate fact of Consciousness is provable in the sum
of all its threefold verifications. II. iii. 26–28, 37–40.
Consciousness, empirically true, and of necessity realized
in philosophy as belonging to the normal man univer-
sally, but heretofore not legitimated, is in each individual
purely and entirely personal. It is his own; and that
which is one's knowing, thinking, doing, loving, may not
be a fact of another's thinking, doing, loving. As special
facts, they certainly are not; each is his own; but each
may have the elements of Self out of which similar
facts may arise. Yet, men are so organized that one has
an excess of actuous force, and a defect of intellectivity
and of susceptibility, and so through these three characteristic elements; the man of actuation cannot comprehend the man of susceptibility, and if the man of intellectivity, without any combination of the other two
correlates, — if such an intellectual monster can be conceived, — can perceive that each of the others is guided, or
rather actuated or inflamed by their respective forces of
character, he can get only the positive fact that they are
actuated by some element or force, which may be given
to him in some formula of words or form of thought,
but it will be void of any content to him; it will have
for him no living, infecundating life. While Consciousness is thus realized as the empirical fact of all normal
psychical life, it is yet susceptible of philosophic proof in
the objective position, which it can hold in and towards
each of its threefold conscious elements. It can say I
do, and it has the knowledge of doing; I think, and it
has the knowledge of thinking; I love, and it has the
knowledge of loving; and it combines all these in its
normal acts, in a positive, if not perfect knowledge and
action of its triplicate life.

15. In the processes of communication from one self

to another self, it must not be overlooked that the self
of the higher and more complete personality may have
much which it cannot communicate to an inferior self
unless in an expanding capacity for progress in such inferior. The inferior has not advanced into that state of
mentalization, or growth of self-normalation, in which it
can intuscept — realize the life of the higher life. How
can he who is full of the light of intellect, illumine the
dark places and narrow chambers of unmentalized souls?
how can the life full of love inweave and impart into the
dry, hard, and sterile nature of the loveless man, except
by the " coal of fire " from the altar of Love, or by
ploughing up that nature with the ploughshare of many
griefs and dependent hopes and personal wrongs, committed by the rude man himself, requiring self-analysis
and self-forgiveness, and making him wish for the love in
mercy, which he had refused? Paul going to Damascus ;
the cruel bigot, or ruler everywhere.   Let it not be for-
gotten that the suffering and sorrow which improves and
purifies and elevates is subjective — it is that which we
bear in meekness and moral obedience ; and it is not the
objective suffering and sorrow — it is not that which we
inflict, and which brutalizes and degrades the inflictor.
If moral culture were an easy process, the meek and
gentle ministers of God would have a more successful
duty to perform, and one, thus filled with love, of more
unction.   But the higher processes of life-integration,
instituted in the trials and disciplines of life, must yet
drive their ploughshares over prelate, priest, and lay-
man. That the Love and Thought, and their *actuation*
into the currents of life by Christ, as a life, is indifferently
intuscepted by the teachers of mankind, ·nay, that in
times of madness they only exercise their animalistic
and human orgasmic forces, is written in the blood of

centuries and continents, and such must be the solemn judgment of those standing aloof from the wretched and criminal movements of political and military life, and who have made the ascent above them, and have seen the *love* of God made the watchword for ruin and desolation. It is easy for bad men in the rostrums and pulpits and elsewhere, from their depraved natures, or mistaken men, in their *mis*-normalated and unmeekly loves, to stir and move the animalistic and human orgasms in mobs and popular masses; and thus it is that even when "pearls" of the highest price are *cast* to them, they turn and rend the giver by scourges, barbarities, and crucifixion, as the old priesthood sacrificed Christ.

16. It is necessary to know the actor and thinker and lover making this intusception. The extent and character of his knowing will be, and only can be commensurate with his entire activities, organic capacities, and the range and qualities of objects knowable. It is, then, necessary to analyze the objects knowable, synthesize them in their appropriate correlations and as they stand in their system of correlations around this central Self. In omniscience all objects and subjects are known; in limited intellectivities nothing is known beyond what has concrete intelligibilities inwoven therein, and as these are correlated to such intellectivities in their organic instrumentalities, as color to the seeing eye, odors to the smell, thought to the thinking mind, volitions to the willing agent, affections to the loving heart. This will be manifested in the progress to the higher ideation of the coördinates in Deity.

17. In making the exposition of the Personality, as comprehended in the whole organization of man, the general phenomena considered are those common to the whole race, and therefore based on the common consciousness

of ordinary men : yet, as indicated in § 15, there are phenomena manifest to some which are wholly incognizable
to others ; so there are phenomena belonging to some
who have been brought into intusceptive communion
with the spiritual life which others have not attained ;
and the truths of both classes must take their places in
any well-considered system of philosophy, on the same
verity of consciousness which establishes the ordinary
facts of psychical or mental life, for it is only the consciousness of each confirmed by the testimony of different classes, and the mutual intusceptions of each class.
In the two latter cases the higher culture of the intellective class and the serener self-consciousness of the morally
spiritual life, and frequently with the very highest intelligence in the latter, may well be put against any negative doubts. In the unfolding process it will be seen
that the ideation of God is a gradual normalation from
the whole elements of nature and life, and these as applied to the Self in a self-analysis, I. i. 8–10, 17–20, 36–
37, 42–43 ; and this even when it is given in a positive
revelation, for a positive revelation gives nothing to those
who cannot perceive, in other words, who cannot give a
life-content from themselves. Hence it has so frequently
happened that the actual revelation communicated by
evangelic men to savage or barbaric races becomes
merged in the surrounding barbarism as soon as the *controlling* life is withdrawn from them. Subordination,
prescribed forms, and continued impressment are necessary
to moral growth in the life of a race. It has always been
given to different men and races in higher forms and
vitality than their native ability could eliminate by direct
personal processes ; yet a moral intonement of life is
sometimes given to those of much limited intellectivity.

18. The content of the solidaric Self, when seized in

its simple discrete elements, their fundamental union, their correlations to life and nature, and their native superintendencies to God, as manifested in so many forms of superstition and natural religion, it is proposed and believed will give a triplicity in unity, and indicate the actual, if not the essential origin of the Solidarity. This view is elemental — it is foundational in man, and man made in the image, and as he ascends, is seen to be *renewed* in the likeness of God, in the exercise of the divine spiritual forces. God knows — Man learns. As man ascends and extricates himself — is extricated — from the cleaving impurities of his animalistic appetites and human appetencies, he gathers clearer ideations of the power, wisdom, and love, which are necessities of thought to any worthy conception of God. Pursuing the Method indicated, but after the manner of strict Rationalism, yet including the whole elements evolved, the Self will arrive at the philosophic coördinations in their unity. To give this philosophic formula a Personality, and make it the living God of the universe, it must have a life-content, in moral correlations with the children of men ; it must be seen as "the *life* which is the light of men," "the light which lighteth every man that cometh into the world, who was and *is* in the world, and the world was made by him" and the world knows him not, and to those who receive Him he shall give the power to become the sons of God. He must be seen as descending to man — and man must be seen as ascending to him on this ladder of power, wisdom, and love.

19. Yet, to make this ideation of Personality more complete, in thus evolving and adjusting the correlations between the triplicate solidaric powers and the trinal coördinations found in the initiate causations, that other relation growing out of these root-forces, and necessary to

the subordination of those individuals and races who are
in their animalistic and mere human conditions, must be
intuscepted in the entire complement of man's nature, in
which the actual and moral necessity of Command as the
effect of Love in the Law-giver is seen as provident,
or wise, loving, and instrumental to aid man, in his
savage and barbaric conditions, in attaining his end of a
purified life of active love, and which, in the incertitude
of human reason and pruriencies of human passions, appetites, desires, &c., contributes essentially to its attainment, which few or none of mankind attain, when
left to the cold and lifeless formalities of the miscalled
• Eternal Reason, I. i. 25, 35–37, as a mere rule of conduct obtained from the light of nature, and informed —
enlifed with the animalistic, or the human loves. But
when Command has inwoven with it the wisdom and the
power of discipline, in love, and applies the exact discipline to the exact disobedience, it is seen that Command
is but another name and form of the Law of Nature
which imposes the discipline of the life and the ages, and
this by the interweaving correlations which subsist thus
between physical nature in all its forms and the bodily,
psychical, and spiritual natures of man. Love in the
Commands is always tendering the Conciliation, when
the subject yields up the disposition to offend. I. vi.
35, 36, 37. Certain physical forces — those farthest removed from immediate correlations with the human Self
— vindicate themselves most openly and conspicuously :
the tree will fall and kill ; and the blow of the murderer,
and the shot and the steel of the murderous battle-
field, will destroy. As the correlations become more
immediate and closely interwoven, they become more
recondite, and sharper investigation and inspection become necessary to see their reciprocal interactions : dis-

eases, as agents of the physical forces, invade the body, and the pathogenetic effects of poisonous and medicinal agents as correlated to various viscera and organisms and the orgasmic forces of life are eventually traced and discovered, and their intimate correlations are understood and established, and used and abused. And in the unfolding of these correlations, the moral forces, from on the spiritual side of the Self, inosculating and interweaving with all the forces in nature and life, are understood and established, and used and abused in the moral movements of life. As these correlations depart from the more open and manifest contrast and open intercommunication between the Self and these correlated forces of nature, the laws of their interaction become more hidden, and slowly it is seen how, in the movements of moral life, interventions, ceremonials, and· commanded observances, and institutions of moral forms, having a life-content of discipline and education inwoven in them, become necessary, and Command is legitimated. Therefore let not the man who is so far mentalized that he catches these correlations with less difficulty of analysis and synthesis, heedlessly destroy the useful forms by which his younger brother is ascending to the Light and Love and Actuation of the higher ‛ moral Life. They are the ladder by which he himself, in his solidaric life, too, has ascended, yea, and has yet to ascend, — only in other and higher forms. The one is germinal, the other is *normalatively* developing. Love, therefore, in the Command is always tendering the conciliation at every step upwards in life, when the object — subject of the command yields the disposition to offend; and the Command to the younger brother, and the accumulated means of knowledge and of action in the elder, control, subordinate, and subjugate the animalistic and human orgasms of

offence, and man in his own actuating spiritual love becomes liege, loyal, and active *in and for* the very Love
which has commanded, and in this very progress has forgiven so much in the very bestowal, yet by the conscious
appropriation by the Self, of this higher Love. Love
repays all that God can give, or man receive. In this
Power, Wisdom, and Love of a Divine father, thus unfolding his power, wisdom, and love in these punitive
retributions of physical, vital, and moral agencies, and in
the authoritative disciplines and instructions of Commands, leading to a higher life, man will perceive his
duties more clearly, and will gain — gains in the processes,
· the Love, the Wisdom, and the actuating Power to become the Son of God. The circle begins in Love, and
ends in Love. The spiritual life below and the spiritual
life above are concordant, and order is justice, and justice
is Righteousness, — and order, justice, and Righteousness
are at one. I. i. 3–5, 15, 16, 17, 34–37, 42 ; §§ 2, 3, *post,*
c. vii.

BOOK FIRST.

THE WORKMAN AND HIS WORK.

———◆———

CHAPTER THIRD.

SUBJECTIVITIES ; OBJECTIVITIES ; AND THEIR SYSTEM OF CORRELATIONS.

1. BEFORE anything can be *determinately* posited in space, its place in space, and its time of appearance, both in relation to things which must and are to appear, must be determinately assigned. Thus the first movement in creation is an intellectual, a moral movement, and it is determinate. The position of a planet of a given volume and density, in a given star-system, must have its appropriated place in such system, I. i. 25; the mass and weight of each planet must be assigned to the place which it is to occupy in the system ; and according as the planet is nearer to, or remote from the sun, as centre of the system, must be the organizations of the vegetal and animal existences which are to appear upon, and inhabit such planet or the respective planets of the system. And before God could have made the autochthonic man of the dust of the earth, he must have made the dust of the earth. Thus moral laws and moral forces enter into the processes of creation from the beginning,

and from before the beginning, and thus are seen to accompany the processes to the end. I. i. 8–10, 41–43; ii. 12–18. Without giving philosophic reasons, just now, this is seen in the differences of vegetal and animal life at different latitudes on the earth. So in the wide sweep of the systems, the same adjusting and balancing forces, which are necessary to keep this planetary system in its harmonious order, are necessary for all the systems, swaying in the boundlessness around the unknown centre. I. i. 10. By positing in space, *place* is made, and space has received an objectivity in it, to which all things coming thereafter have relations of time and space. I. i. 7.
• Before a cognitive agent, organized as man, is posited in place, there must be an objective standpoint — something standing under him to occupy; and objects to cognize, to use, and abuse, and correlations are predeterminately necessary. The creation of this standpoint, this theatre, requires such a creature as man in his aggregate of humanity, in his autonomic functions and spiritual powers, to give a meaning and a life to the whole of creation.

2. The power of outward, objectifying action, of intellectualizing, and of loving, either in beastliness of gratifications or in charities of life, may differ in degrees, in intensities and capacities, but they do not differ in kind, of their respective kinds. Each may be directed towards various objects, as they evidently are; but at base it is but one Actuating Power, which goes out into action; it is but one Intellectivity that thinks, arranges, devises, plans, fixes time and place for action, &c.; and it is but one Love in manifold forms, means, and objects of gratifications. By that law of thinking in the subjective Self, however derived, by which the uniform phenomena constantly appearing in or arising out of one thing are

referred to their common substance, I. i. 10, these respective powers must be seen as arising out of their respective sources of causation.

As the proleptic steps of the movement of nature are traced back to a beginning, the secondary causes are seen to converge in to the Initiate Causations. I. i. 8–11. As these secondary causes are thus retraced from their complex combinations in nature and life to the simple bases from which the *whole* originated,· new differences of functionalizations must be seen at each step of the progress as inwoven by and from the Initiate Causative Forces. The difference and the permanence inwoven in these new forms of forces, so as to create classes, orders, families, and species, are seen in the vegetal and animal kingdoms. Science is so accepting the facts.

" One important truth already assumes great significance in the history of the growth of animals; namely, that, whatever the changes may be through which an animal passes, and however different the aspect of these phases at successive periods may appear, *they are always limited by the character of the type to which the animal belongs, and never pass that boundary.* Thus, the Radiate begins life with characters peculiar to Radiates, and ends it without assuming any feature of a higher type. The Mollusk starts with a character essentially its own, in no way related to the Radiates, and never shows the least tendency to deviate from it, either in the direction of the Articulate or the Vertebrate types. This is equally true of the Articulates. At no stage of growth are their young homologous to those of Mollusks or Radiates, any more than to those of Vertebrates, and in their final development they stand equally isolated from all others. That this is emphatically true of the Vertebrates has already been fully recognized ; and the facts known with refer-

ence to this highest type of the animal kingdom might
have served as a warning against the loose statements
still current concerning the so-called infusorial condition
of the young Invertebrates.   These results are of the
highest importance at this moment, when men of authority
in science are attempting to renew the theory of a general
transmutation of all animals of the higher types out of
the lower ones.   If such views are ever to deserve
serious consideration, and be accepted as involving a
scientific principle, it will only be when their supporters
shall have shown that the fundamental plans of structure
characteristic of the primary groups of the animal kingdom are transmutable, or pass into one another, and that
their different modes of development may lead from one
to the other.   Thus far Embryology has not recorded
one fact on which to base such doctrines." — Agassiz's
Methods, ch. xvi.

" Embryological investigations have taught us that
during the incipient growth of the higher animals none
of *their* organs exist, and yet the principle of life is active,
and even after the organs are formed they cannot act at
once, most of them being enclosed in the whole structure,
in a way which interferes with their later functions.   In
the little chicken, for instance, the lungs cannot breathe,
for they are surrounded by a fluid ; the senses are inactive,
for they receive no impressions (?) from without ; and all
those functions establishing its relations with the external
world lie dormant, for as yet they are not needed.   But
the organs are there, though, as we have seen in the
turtle's egg, they were not there at the beginning.   How,
then, are they formed ?   We may answer that the first
function of every organ is to make *itself*.   The building
material is, as it were, provided . . . . Before the lungs
breathe, they make themselves ; before the stomach digests,

it makes itself; before the organs of the senses act, they make themselves ; before the brain thinks, it makes itself. In a word, before the whole system works, it makes itself out of the elements given by the formation of independent eggs (ovarian egg) ; its first office is its self-structure."— Agassiz's *Meth.* ch. xv. This is true and it is not true. The function of thinking as a determinate process of thought in the formative production of the organisms of the incipient brain is not the function of these organisms after they are made. So of the various Loves, — and so of the executive power of doing, in so many forms of doing. The thinking, the loving, and the doing are the functions of the organization *after* it is substantially completed. The various forms of these manifestations as seen in the instincts of animals and men, in the spontaneities of man and in his self-conscious conduct, in certain of their forms and natures, proceed from and in others are exercised through adapted organisms. The explosive passions and the attracting loves are experienced in various forms of spontaneities indicating their different organic functionalizations. The forces of functionalization, which builded the organs and endowed them, preceded the respective organisms. The autonomic egg — the primitive germ, including, potentially, the whole of the creature — preceded the structure of the parts. The functions of the brain did not build and endow the brain in its variety of functions, but the intelligential and intelligible forces assigned to each autonomy, as a general whole, was precedently necessary to the structure of the parts. There are functions for growth, and there are functions for action after growth. The general autonomic forces of the type, in each species, include, govern, and mould the included special forces of each subsidiary organism, as they, in turn, affect the general autonomic type of each individual. In man, the solidaric autopsy is superimposed, and the function of conscious self-normalated conduct is reflex, — is an inverse action from the consciously objective position of the Self controlling and normalating the various functions of these various organisms.

3. The vegetables build their respective differentiated forms with an accurate *intelligence* inwoven into their respective autonomic germs, which, in formative production of germs, preserves and continues each *kind* after kind; in animal autonomies the special differentiated forms of orgasms subservient to the whole organization of each autonomy, with accurate intelligence, greatly diversified for each, build each separate organism, and these as subordinate to the more comprehending intelligence, which weaves the characteristic *form* of the particular species, which thus comprehends and includes the special orgasms and their organisms. The special intelligential forces, which weave each special organ, are subordinated to the intelligential force, which gives the whole complete Form of the thing. The human autonomy ascends in its constructive intelligence and subordinates the entire variety of the functionalizing forces, which so wisely build each separate part, and gives distinctive form to man. The subordinate forces converge, under his autonomic germ-forces, to the production, preservation, and continuance of his autonomic species. All these organisms are modes for the action and reaction of forces, which is signified in the popular language of "the wear and tear of life." This is becoming, has become, the well-settled law of science. Carp. H. P. § 345, says: "Thus it comes to pass, that, during the whole period of active life, a demand for nutrition is created *by the very exertion of the vital powers*, but more especially by the evolution of

the Nervous and Muscular forces. The production and application of these, indeed, may be considered as the great end and aim of the Human organism, so far at least as the individual is concerned." And the evolution will show, that all these are positively and directly affected by the activities, passions, mental labors, and emotions of the Self.

4. These various autonomies, vegetal, animal, and human, are, more or less, in the philosophical contingencies, dependent on a supply and modulation of forces, from the various plasticities of nature, and on the assimilation of forces prepared by the inferior organisms on which some of them subsist, and upon the geotic causes, as modifying forces, with which they may be surrounded according to their longitudinal, latitudinal, and isothermal lines, and soil, and other local influences, and in man especially, on civilization and evangelizations, and these frequently acting with antagonizing efficiencies. I. i. 13, 26; ii. 11, 15. Each planet, and each place on the planet, will, therefore, give a peculiar modification to each autonomy, while the autonomy will hold the entire organization, under these certain limitations, true to the form-giving forces in its type-autonomy.

5. Man is thus given in his Solidarity and his Autonomy. The Autonomies furnish the organisms of Sensation to the Self, for the cognition of external things in their differences of quantities and qualities, and orgasms for the impulsions of the animalistic organisms, and to give modality of forces and direction to the human passions, intellections, and emotions, and for the means of reaction from the Self — the Spirit, in, on, and through, and over these orgasms and organisms. As is the autonomic organ, so will be the seeing; as is the ear, so the hearing; as the erotic desire, as the organs of hunger and thirst, of cun-

ning, of ferocity, of covetousness, &c., so the impulsion in the inner organization of the animal or the human Self. But while this is so, and as the eye, so the sight, but not the painter-artists within; as the ear, so the hearing, but not the musicians within; and psychical subsistences with their inherent orgasmic forces are necessarily notionalized as within, I. ii. 6–12, in their differentiations of special forces and diversified organisms as necessary to the manifestation of the artist, of whatever kind — painter, musician, poet, thinker, and doing-thinker, and thinking-doing-lover, as the solidaric Self of each self.

All living animals have nerves. All of the more highly organized animals, fish, reptiles, birds, quadrupeds, Man, have Brain. And the variety of Powers, or Diversification of functions, corresponds to variety and complexity in the organization of the whole nervous, vascular, and muscular systems, thus complementing a body of correspondent parts. And the nervous system, in the higher organizations, is observantly seen to divide into systems of sensitive — afferent nerves carrying in sensations and sensibility, and of motor — efferent nerves carrying out power or force for motion, and the sympathetic nerves, which perform independent functions of their own, and keep up certain kinds of communications between important viscera and the cranial organization.

In the human subject, the whole encephalic mass, the Brain, is composed of eight distinct parts: — 1. The Hemispheres, or the cerebral ganglia, constitute about nine-tenths of the whole mass of the brain; this Cerebrum occupies the upper cavity of the skull, and is by universal consent of scientific men allotted chiefly to *mental* powers; and, as a general rule, in healthy conditions, mental powers in different races and in different

individuals correspond to the size of the cerebrum, it
will further become apparent, that the due combination
of Active, Intellective, and Affective powers are alone
properly called Moral Powers. 2. The Cerebellum, with
its transverse convolutions, situated behind the cerebrum,
to which, by some, have been assigned organs of instinct
— the impulsive spontaneities of the human passions and
affections, but which by others is supposed to be the seat
of combination of organs, or instrumentalities for coördinating the powers or positive forces, by which the different voluntary motions are regulated and harmonized, so
as to produce order and determinateness in the movements of man. 3. The Tubercula Quadrigemina, which
give rise to the optic nerves, and preside as ganglia over
the Sight. 4. The Tuber Annulare near the centre of
the brain, and intimately connected with Sensation and
voluntary motion. 5. The Olfactory ganglia give rise
to the special sense of Smell. 6. and 7. The Corpora
Striata and the Optic Thalami, small bodies, whose functions are undetermined. 8. The Medulla Oblongata,
long known to be particularly essential to the preservation of life, so that it has received the name of the "vital
point," the "vital knot." In immediate connection with
this last, and in immediate or mediate connection with
all these, are the Spinal Cord and the system of the
Great Sympathetic Nerve. The former is but the prolongation of the Medulla Oblongata, and is a long ganglion covered with longitudinal bundles of nervous filaments, and is the "Silver Cord," occupying the central
cavity of the backbone, and which sends out nerves to
each successive portion of the body, and supplies with
nerves the muscles and integuments of the whole body,
which are endowed respectively with the power of conveying sensations and communicating motion, depending

on the integrity and activity of the Spinal Cord. The Spinal Cord is also in direct connection and communication with the system of the Great Sympathetic Nerve, which consists of a double chain of nervous ganglia running from the eyes to the lower extremity of the body, along the front and sides of the backbone, and connected with each other by slender filaments, as they are also direct in communication with the Spinal Cord by motor and sensitive filaments received from it at different important points. Branches from this Great Sympathetic Chain are distributed to organs over which the "consciousness and the will" have no immediate control, as the intestine, kidney, and heart, &c. Through this Spinal Cord, the Cranial Nerves, and the system of the Great Sympathetic, all in immediate or mediate communication with these divisions of the encephalic masses, and with some Autopsic Power presiding in and over them, all sensibility and sensation inwardly, and movement outwardly, in and from the human frame, are communicated by specific bundles of these nervous filaments. The whole taken together in the whole, and in the actual working correlations of the parts, and in the distinct function which each part performs in itself, yet always with reference to the whole, show, throughout, a division and separateness of organisms, functions, and functionalized forces in an organization of a complete whole for the action of autonomic forces, for spontaneous instinctive and psychical actions, for intercommunication of action and reaction, and for the determinate interposition of action by the Autopsic Self — and the fact of Form and Functionalized Forces prevail everywhere, and their mutuality, and their dependence on the controlling type which includes all. The fact, the extent, the increase of diversified functions, and their uses in each separate part of these respective divisions, and their communicating filaments, and their appropriateness in the combinations of the whole, will be readily and more fully comprehended by a slight study of Comparative Anatomy and Comparative Physiology; and as the series shall unfold and ascend from the lower to the highest, it will be recognized how provisions have been made in the unfolding ascent for the introduction of higher and still higher powers, until there are distinct, yet correlate provisions for the distinct and harmonious action of the Active, Intellective, and Sympathetic Powers of Man as a determinate, autopsic Moral Power.— Dalton's *H. P.* § ii.; Carpenter's *H. P.*; Id. *Comp. P.*; and I. i. 1, *ante.*

6. Man, then, has a threefold life : *a.* Somatic life ; a life which belongs to the bodily organization, which has, more or less, something in common with all animate existences, and which acts concordantly and subordinately to the inclusive autonomy, which separates him from all other existences, and gives him his form and species as man. This is his comprehensive, formative, differentiated germ-power as a human creature. I. i. 25–33. *b.* The interior soul-life; a psychical organization so constructed as capable of acting by orgasmic organisms on this bodily organization, and through it on nature and in life, and from one orgasmic organism on another, all of which organisms are correlated to the spiritual Self, so as to send in sensations and impulsions to this inner Self, and receive direction, control, and subordination from it. These soul-orgasms write their special influences on the different correlated organisms of the body, as shame on the cheek, fright on the sphincter muscles, awe on the hair of the head and over the body ; there are bowels of mercy, &c. It is seen as that psychical organization through which the proper Self can select this set

of muscles instead of that, through some special convolute of the brain, not weighing many grains, and in direct communication with the particular muscle or set of muscles, and determinately send forth motion-force by the special muscle, and lift — overcome resisting weight — static or dynamic force. This is the life, which may become automatic as in monomanias, I. i. 31, which displays its violence in diseased conditions of the organisms, and which so constantly exhibits the peculiar characterizing quality in those individuals who are driven, uncontrollably, by one idea or passion in life. The exacerbations of these orgasmic forces are the sources of hallucinations and fanaticisms in individuals and in society. c. The life which thus receives information by Sensations from without, and from the orgasms within, and communicates outwardly and from one organism to another, thus capable of being ultroneously chargeable with forces autopsically selected for the commission or prevention of its deeds — its own creative facts, is the Spirit's life — the moving powers of the solidaric Self.

7. Life, in its simplest acceptation, is Vital Activity, and obviously, therefore, in organized bodies, involves the notion and fact of change, an exchange of forces. I. i. 8–10. The life of any complex organization, as of the somatic, and somatic and psychic organisms of man, is in fact the aggregate of the vital activity of all its component parts ; and no fact has been more clearly ascertained by physiological research than that each elementarily organic part of the fabric has its own *quasi* independent power of growth, that it has its own existence, that it goes through its own sequences of vital actions, in virtue of the endowments of forces, inwoven in the forms and substances which evolve them, and of the influences, as forces also, to which it is subjected during the progress

of its existence. I. i. 3, 5, 13, 24. But in every living
structure of a complex nature there are a great variety
of activities, resulting from the exercise of the different
powers of the several component parts ; yet there is a
certain harmony or coördination amongst them all, whereby they are all made to concur in the life of the organization as a Whole ; that is, there is a congeries of *quasi*
independent organisms, each performing its blindly intelligential, and therefore intelligible function, yet these
functions are subordinate to a higher blindly intelligential
function inwoven in the germ-autonomy, which controls,
converges, and uses them all to produce, preserve, and
continue the specific species of each.

If the history of the Life of a Plant is contemplated,
it is seen to grow from a germ to its particular fabric, in
virtue of its subordinated organisms as located in its
roots, trunk, branches, leaves, blossoms, and fruit, and
the plant is sometimes of gigantic size. Not only so; it
not only generates a large quantity of organized structure
constitutive of its actual growth and productive of its
fruits, but frequently produces many organic compounds
from the plasticities of its *locale* which do not undergo
organization, as the various secretions of plants, trees,
and even of some animals, as gums, musk, &c. By
a *subsexual* intercourse, plants multiply their species
through germs. So far there is no conscious willing,
doing, thinking, or feeling ; yet the work is done as curi-
ously intelligential throughout as if it were conducted in
each plant, and in each organism by a consciously intelligent superintending autopsy. The life-forces are there.

In analyzing the operations of Life, which take place
in the Animal body, the difference of organisms and
their comprehension in one inclusive autonomic organization are essentially the same, and differ only in the con-

ditions under which they are performed, and, through the
zoöphyte, the Plant, and the Animal, blend and mingle
into one system of onward movement, yet in orders of
diversification, in which vegetal life is seen as the homologue of animal life, and all rising up out of fundamental
Forces of Life. In vegetal life the plastic forces and
their concomitant substances, I. ii. 6, which contribute to
their growth and their organic secretions, are taken up
more immediately from the plasticities in their simple
unorganized conditions; in animal life the plastic forces,
which contribute to the growth and action of the different
organisms, and their inclusive autonomy, are taken up
from the inferior and subordinate vegetal lives, and from
the plasticities in their simple unorganized states, as the
animals feed chiefly on vegetal tissues, and some on these
and animal tissues, and all on air, water, light, &c. And
thus the animal autonomy maintains the integrity of its
parts, by virtue of assimilations from the vegetal forces
and the simple plasticities, against the destructive effects
and consequent exhaustion of and by its various activities. Food is necessary to repair exhaustion and supply growth. This is performed in the animal economy
by digestion, assimilation, and circulation. In the exhibition of psychical forces the change and alteration of
brain-tissue is supplied by forces taken up from the
autonomic into the psychical uses. In man a larger
proportion of blood goes to the head than to the body.
The chief functions of the animal autonomy are so bound
up together that digestion, assimilation, circulation, or
excretion cannot be suspended without the cessation of
all; if one is destroyed, all are involved. The properties
of all the tissues and organisms are dependent on their
regular Nutrition by a due supply of duly elaborated
forces furnished from the proper food, air, water, &c., in

and through the blood; and these cannot be supplied, unless circulation, respiration, and excretion are duly performed. The homology of vegetal to animal life may be seen in those cases of animal life where a human creature is deprived, as in apoplectic coma, of all the powers of animal life, sensation, voluntary action, &c., and yet is capable of maintaining a vegetative existence, in which all these parts of the organic functions go on as usual; "and a similar experiment is sometimes made by nature for the physiologist, in the production of fœtuses, as well of the human as of the other species, in which the brain is absent and which yet can breathe, suck, and swallow, and perform all their organic functions." — Carp., *H. P.*, § 25. The first life as herein stated in section five is exhibited.

8. "A change of the composition of the organized fabric in man is a necessary condition of every manifestation of its Vital Activity; it is therefore requisite that provision should exist for the replacement, by new matter having new forces involved therein, of all those particles which, having lost their vital endowments, are in process of return to the condition of inorganic matter, yet which in themselves possess or become charged with new forces for other organizations. Hence, of course, every increase of the activity of the animal functions becomes a source of augmented demand for nourishment. A constant supply of Aliment is therefore needed for the maintenance of the body in its growth, and after its full development for the supply of these exhausted forces." — Id. § 26. It may now be seen by the observant eye, that, as in section five and *ante*, the somatic life is wholly dependent on its combination of organisms — and this even a truncated one, there are psychical forces of the higher organization, which, while they are dependent

on nutrition, &c., yet derive strength, versatility, and expansiveness from forces directed from within outwardly, and are repercussed from one organism to another. This is the psychical or Second Life mentioned in section five.

"Every part of the bodily frame circulates more or less rapidly. 'At every moment,' says Liebig, 'with every expiration, parts of the body are removed, and are emitted into the atmosphere.' The motion of one part of the body involves the motion of every other part. The mechanism of certain parts admits of action more instantaneous than the quickest suggestion of the *will*."

"The power of volition with which man is endowed is never allowed to rest; for he finds himself constantly *solicited* by different objects or attempting to master the difficulties which lie in his path" (to some soliciting object). "If the difficulty relate to an object of knowledge, spontaneously the mind tasks its powers to pierce the obscurity, and this effort is 'a concentration upon a point of forces before diffused.'" "Man is a cause, and is constantly acting under the conviction, that, amidst all the external influences which surround him, he has the power of *reaction* and self-regulation." "Without object or impulse every part of our active nature would be soon lost to us, or rather would never be known to us. But with these that active power is disclosed to us; by exercise it is increased; difficult and occasional acts become easy and confirmed habits; physical weakness is" (not unfrequently) "replaced by muscular strength; ignorance by knowledge" and versatility and certainty of intellectual power; "and a sense of duty grows into a course of intelligent and delighted obedience. Thus activity is a law of our nature, and the condition of its development." — *Man Primeval*, P. I. c. vi. §§ 2, 6, 7.

"We have referred to the great truth that Force, like

matter, is persistent and indestructible. Its changes are but mutations from *form to form;* an impulse of force can no more be created or destroyed than a particle of matter. This principle is known as the *conservation of force,* and is characterized by Dr. Faraday as the highest law in physical science which our faculties permit us to perceive." —Youman's *Class-book of Chem.* § 402. Passing over this author's want of perception of the finality in these processes, that matter is but *forms* of forces in stabilitation or action, further views and facts are selected from the same work, §§ 1261, '62, '63. He says : " The amount of thermal force generated annually in the body of an adult man is sufficient to raise from 25,000 to 30,000 lbs. of water from the freezing to the boiling point. All the acts of the body, every motion, every utterance, breath, or thought, *consumes force.* We make about 9,000,000 separate motions of breathing in a year ; thereby inhaling and expelling 700,000 gallons of air. At the same time the heart contracts and dilates 40,000,000 times — each time with an estimated force of 13 lbs., while thousands of tons of blood are annually driven through the heart and general system. Besides these involuntary acts, the organism generates force for a thousand forms of voluntary physical action. A healthy laborer is assumed to be able to exert a force equal to raising the weight of his body through 10,000 feet in a day."

" Corresponding to this activity is a high rate of internal change. The living body is like a waterfall : while it appears an unvarying form, it is yet composed of particles in swift transition. A man consumes in a year about 800 lbs. of solid food, the same amount of oxygen, and about 1500 lbs. of water, or altogether a ton and a half of matter. Chossat ascertained the waste in various animals to be an average of $\frac{1}{24}$ of their weight daily ;

and Schmidt determined it to be, in the case of a human being, $\frac{1}{23}$ of the weight. Johnston says an animal when fasting will lose from $\frac{1}{14}$ to $\frac{1}{12}$ of its whole weight in 24 hours."

" In the exercise of functional power, parts waste and are ever renewed. In all the deepest recesses of the body, in every muscle and conducting nerve, and even in the thinking brain, myriads of atoms are constantly dying and being replaced. As soon as we begin to live and act, we begin to die. The decomposition is in proportion to the activity. Muscles are rapidly changed, and are always more or less acid from the oxydized products in their substance. It has been fully proved by G. von Liebig, that muscles absorb oxygen and exhale carbonic acid as long as their contractility lasts. With the exercise of a muscle, blood is urged toward it; if the current is stopped, it is paralyzed. So also with the nervous sys tem ; brain-power is dependent upon cerebral transforma tions. Indeed, changes go forward more rapidly in the brain than in any other part." " So rapid are its trans- formations, that, though but $\frac{1}{30}$ the weight of the body, it receives from $\frac{1}{5}$ to $\frac{1}{10}$ of all the blood driven from the heart, to maintain its normal waste and repair." — Id. § 1268.

These views are selected not for the special facts which they present, but for the general fact which they involve, that nature and life are but a mutation of forces from forms to forms, yet that there are underlying Forces for muta- tion. And in the general and indefinite use of the term forms, the fact must not be overlooked that the produc- tion of forms, and the action of forces in each specific form, are in virtue of Powers higher than and dominating the general forces capable of acting under such varied ordinate direction, and in the organic forms of such mul- tiplied diversity. In each individual vegetal and animal

life there is a diversity of organisms, and in the more
highly instinctive or intelligential creatures the diversity
of organisms and functions increases. Yet this diversity
of organisms and functions is subordinate to the autonomic forces which make each species, vegetal and animal, what it is.

And again; these facts of the use, action, and reaction
of forces thus inwoven in the human system, show,
that, as the forces are exercised, controlled, and used, or
abused, so will the organization be more or less effect-
ively changed and moulded. And so the man becomes
the diaphanous ectype of the inner spiritual self, as he is
moulded by geotic and extraneous moral causes, and as
he moulds his surrounding organisms, from instant to
instant, in his animalistic propensities, his human desires
and purposes, and in his higher spiritual manifestations
of autopsic actuation, intellectualizing and loving. I. i. 1.
Yet all changes are subordinate to the persistent form of
the type in the separate individual, and in the generations in their species.

9. Self-conscious Thinking, Loving, Doing, depend
upon an organization for their manifestation, outwardly
towards nature and life, and inwardly in acting and reacting on and from one to the other for such manifestation in human differences of character. These, thinking, loving, and doing, proceed from a centre where
Consciousness, common to the three, notes their difference
and their normal union. This is the third, the Spiritual
Life. I. i. 34. But here examine *yourself*, as well and
closely as your habits of self-analytic reflectiveness will
enable you. Anger, wrath, sudden self-defence, are seen
as spontaneities. In the child, and the uncultured and
unnormalated, impulsive man, the blow, the rough, rude
word, and the instinctive self-defence, are *projected* at

7

once into spontaneous action. Intellection — reason comes after, and in the regrets of prudence or the penitence of moral feelings shows that this Self would have *restrained* the impulse with more time or other occasions of reflection. In some men the cultivated Intellectivity, with a *love* of prudence, or a *love* of purer life, *habitually* restrains this projectile spontaneity. The fact becomes apparent that it is a force restrainable. The same or very similar processes show that it may be held in abeyance, and after Reason has acted and determined its conduct from a prudential love, or for some gratification base or elevated, animalistic, human, or spiritual, the outward action will become conformed to the impelling or the directing power within, and, in determinate action, to some plan of action or of conduct. In the determinate selection of the efferent motor nerves to be called into action, as this finger or that, this arm or that, for this motion or that, to accomplish this causative end or that, is not only seen the special nerves of motion, but their special and specific insertion into a distinctive organism of the brain, so as thus to be selected in the instinctive repercussion of one organism through another, and also, in determinate action, the special manifestation of the determinated act from the Self is given. Again, observe the Love in all the special gratifications; they, too, are blindly impulsive, but attractive to and around the Self; the animal and the animalistic gratifications impel, by sending the afferent orgasms on each appropriate afferent set of nerves through the efferent nerve, into which it is inserted, § 11, *post,* and so the action is repercussed into overt manifestation. The animalistic orgasms in a man with strong animal organization, move much, not altogether, yet in some instances altogether, like the animal repercussion. So it is in the human desires, hopes, ambitions, covetousness, &c., in the strength of their spontaneous movements, in which the intellective organisms are seen, in like manner, as their mere agents. But the love, in these appetizing impulsions, must act through the intellective organism for its time, place, mode, and means of gratification. So the distinct organisms for the intellective power act in monomania, revery, and some forms of senile insanity, in which they are carried on unconsciously without disturbance to other psychical functions. But the proper Self, in its own clear self-consciousness, rises up above these unnormalated spontaneities and activities, and selects its motive sentiments, and determinates its conduct to the heights above or the depths below. Here some of these functionalized loves of gratification are found widely separated from the brain-organism, as in hunger, the erotic desire, &c., yet are in direct communication with it, while others of them are evidently repercussed from the brain, and manifest themselves in effects on the viscera, as in shame on the cheek, in bowels of mercies, in passions, or depressing emotions on the digestive functions, &c. As there are organs for direct action, so there are organisms for interaction and for the determinate interposition of action. I. i. 17–23 ; ii. 11, 14–19. All these are seen as positive forces in their temporary or transient effects on the animal and the human systems. The child knows the angry or wrathful man by the effect of the passion on the features of the face and in the violence of his muscular system ; and he is attracted by the placid yet genial effects of the good man within. And these causes, continued, write their characterizing effects immanently on the human system. I. i. 1. And observant men generally recognize the reflective man in his appearance. So clearly are these effects referrible to designate causes so identical, or identical with those mentioned, that Dr. J. Mason Good, *Book of Nature*, pp. 431, 432, says: " In contemplating the Passions or other Affections of the mind as cognizable by external characters, they easily resolve themselves into *two* descriptions, the Attractive and the Repulsive, the signs of which are to be sought for in man and the nobler ranks of animals in the face, but considerably also in the attitudes and motions of the body. In the Attractive affections, the features, limbs, and muscles are uniformly soft and pliant; in the Repulsive, as uniformly tense, and for the most part rigid." He probably did not use the words Attractive and Repulsive with any thought of their discriminate underlying forces as specific causations producing their diverse, discriminate effects. It will be seen that all these are controllable and controlled by the autopsic Self, yet in its own independency for some specific love, and by its proper intellective direction for actuation.

Man strikes spontaneously in sudden self-defence or anger. But when the blow is *cool* and determinate, it is intensified with the determinating, the intellective force, and is more powerful. When it is determinate and cool, yet is for the gratification of a deep and pervading Love, it will be seen and felt as more intensified and more powerful. There is a combination of all the Forces. Yet as these forces increase, and the moral power of control and subjugation of these forces increase, there is an ascent and purification of the Moral Life.

10. Thus starting with the three correlate powers in the Self, and beginning with the three initiate causations, I. i. 8–10, 17–23, and following them in their subtle but intelligent and thus intelligible windings through their involutions in the works of nature and their evolutions in active forces and in autopsic life, they will be seen as

objectivities from the coördinate and eternally perduring forces, as they came forth into creative manifestations from the central Personality of the universe, always preserving their intelligible identities and their coördinating essentialities, and which are at no time separate in him, except by that Future-Now in his divine omniscience which assigns an order in the relations of time and space, and in the correlations for the workings of his great Prolepsis. I. i. 6, 42, 43. Standing on the summit of the generations and the geologic eras, the Self must begin with the elements of intelligibility incorporate in himself, and tracing them back to the initiate forces, it can only find its own intelligibilities in the initiate intelligence which moved over into *objective* creation and established the intelligential order of on-going which culminated in the emergence of its own clear autopsy. So with his power of conscious overt action — his objectifying from his Self; so with his variated loves; and thus he arrives introvertedly at the coördinate coessentialities. Thus, again, it is seen that it is from some intrinsic and inherent communities of identical natures or consimilarity of the elements at work that intelligent man catches and comprehends the intelligibilities inwoven in nature and life. To refer again for authoritative sanction that these processes are neither too bold nor too low, the authority of that work having upon it the imprimatur of that Church which is most exclusive in its forms, dogmatic in its instructions, and slowly careful, and conservative in its philosophy, Cortes, *id*. B. II. c. vi., is quoted: " All things had come forth from God, and were to reascend to God, as to their first principle and origin; and because all things were created by him and were to return to him, so was there nothing that did not *reflect* with more or less brightness his beauty. . . . . For this reason the universe, which signifies everything created by God, *is the combination of all substance;* and order, which signifies the *form* in which God has modelled all things, is the combination of all beauty. There exists no creator except God, there can be no beauty except in order, and no creature except in the universe." But this God as creator, and this order and beauty, can only be fully beheld in the entirety of that Prolepsis as seen from the deific, the transcendental standpoint, from before the beginning and in the verifying fulfilment of its history in the aggregate movement of the worlds and the creatures and the responsible natures he has created — the moving complexures of the universe. I. i. 42, 43; ii. 12–14.

11. This gives the complexus of the Human Personality standing upon and amidst the solid structures, beauties, wonders, and strifes of its planetary theatre. The inner subjective Identity of this autopsic Self is the human Consciousness holding, as in a complexus, the somatic body, the psychical organization, and this its threefold subjective identity. The somatic body is builded up in virtue of its special functions and organs distributed to each part, yet subjected, with more or less of contingency, to the subordinating *form* which makes it a man, and a man of its particular race. These subordinated functionalized organs and organisms are called " ganglionic centres," and are " essentially composed of vesicular substance made up of cells which may be spheroidal, fusiform, caudate, stellate, or of almost any variety of shape ; the latter, nerve-trunks, consisting entirely of nerve-fibres, which, in their most completely developed state, are tubular. All our knowledge of the structure and endowments of these two forms of tissue renders it probable that they bear a complementary relation to each other; the vesicular substance having for its office to *originate*

changes which it is the function of the Fibrous to conduct.
And thus by means of the extensive ramifications of the
nerve-trunks, and the power of instantaneous transmission which they possess, almost every part of the body
is brought into such close relation with the Central Sensorium, that impressions made, even at the points most
remote from it, are immediately felt there, provided the
nervous communication be perfect; while the influence
of Mental States in determining movements " (from inwardly to outwardly) " is exerted no less speedily and
surely upon the muscular apparatus.   For the transmission of these two sets of Impressions — the Centrifugal
and the Centripetal — two distinct sets of Fibres are
provided, *neither of which is capable of taking up the
function of the other;* these are termed respectively the
afferent and *efferent.*" I. ii. 11, 6–8.   "Of the mode in
which the latter terminate *in* the central organs toward
which they pass, and in which the former commence their
course in these same organs, no general statement can
yet be made; but it is quite certain, that, in many instances at least, there is an absolute continuity from one
form of nerve-tissue to the other.   Three principal modes
have been ascertained in which this may occur.   Either
a globular cell may give off a prolongation that becomes
a fibre, in which case the cell is said to be unipolar.   Or
a ganglion-cell presents itself, as it were, in the course of
a nerve-tube, having each of its extremities prolonged
into fibres, in which case the cell is said to be bi-polar.
The former of these arrangements seem to be more
common in the nervous centres of man and the higher
Vertebrata, whilst the latter prevails in fishes.   But in
certain parts of the nervous centres of man we meet
with ganglionic cells sending out radiating prolongations
to the number of three, four, five, six, or more ; some of

which are to be traced into continuity with the axis-cylinders of nerve-tubes, while others, it is probable, in-osculate with those of other stellate cells." — Carp., *H. P.*, § 457.

Thus far for the organisms through which the mobilizing forces are transmitted. Observing that in the systems of afferent and efferent nerves, "neither of which is capable of taking up the function of the other," it will at once be apprehended that the reason of this important fact is, that the structure of these nerve-tissues are so different that the one is incapable of transmitting a force which is proper to the other, or that the force affered, sent in, is different in determinate power from the force effered or carried out, or, what is more probable, that both are differentiate. Rationally considered, this must be so. Yet when it is considered that so very large a portion of the growth and continuous and periodical or occasional action of the different parts of the system are ganglionic, as in the growth and sustentation of the different parts of the system, the pulsation of the heart, the expulsion of fæces, the temporary love of the animal dam for its young, &c., yet that they are necessarily connected, in the life of the whole, to other parts of the system for digestion, assimilation, circulation, and excretion, it may then well be seen that many of these nerves, which originate action, will enter into and continue an unbroken current, through other nerves, the forces necessary for the fulfilment of their respective economies. I. i. 30. This will explain ganglionic growths, and gives, in a higher combination of organisms, the operation of instincts, and brings out more fully the idea and the fact of a higher organization for the local habitation — *locum tenens* — of the autopsic Spirit and for its determinate interposition of action in receiving impulsions and infor-

mation from an afferent nerve, and then, upon determination, using this or that efferent nerve for action or
not using them at all. While this is so, these very considerations show that in human life, when these organisms are too closely inwoven together, or some of the
organs preponderate, the life of such individual is not
an autopsic but an orgasmic life, — the constitutional
thief, the natural murderer, the helpless and hopeless
prostitute, the native buffoon; the crow, the fox, the tiger,
the goat, and the monkey, substantially reappear. I. ii. 12.

That the forces which thus act are peculiar may be
gathered from some considerations of nerve-force presented by Dalton, *H. P.* § II. ch. ii., where he says: " It
will be readily seen that the nervous force or agency by
which the nerve acts upon a muscle and causes its contraction is entirely a peculiar one, and cannot be regarded either as chemical or mechanical in its nature.
The *force* which is exerted by a nerve in a state of
activity is not directly appreciable by any of the senses,
and can be judged only by its effect in causing muscular
contraction. This peculiar vitality of the nerve, or, as it
is sometimes called, the ' nervous force,' does not precisely resemble in its operations any of the known physical forces. It has, however, a partial resemblance, in
some respects, to electricity." . . . " The latter force, to
exert its characteristic effects, must be transmitted through
isolated conductors, so arranged as to form a complete
circuit. . . . Moreover, the nerve, in order to conduct *its
own* peculiar force, must be in a state of complete integrity. If a ligature be applied to it, or if it be pinched
or lacerated, the muscles to which it is distributed are
paralyzed for all voluntary motion, and yet it transmits
the electrical current as readily as before." The more
these two forces are studied, the greater will appear their

fundamental similarity and their differentiate qualities, the one to operate on nature and the other to act by its double mode of action — afferent and efferent — in life. This nervous force is the movement force of all animal, probably of all vegetal life, fulfilling varied and wonderful intelligential offices, — now building up the autonomic growth of the foot, now moving the autonomic pulsations of the heart, now driving the automatic impulsions of the instinctive acts of the animal, now the repercussing agency of the psychical orgasms of the brain in animals and men, and now the conscious forces at the disposal of the ₵elf, controlling these very orgasms and normalating all forces into its system of life. Their subordination indicates the spiritual supereminence of the ₌Self.

Thus it is seen how provision is made for the action and reaction of ganglionic centres, and how they mutually contribute to the somatic and redactive life of the body. And how provision is made for the centripetal transmission of impressions from without and from the animalistic impulsions in the corporeal orgasms to the Self within by means of the afferent nerves ; and that a different organism, to be acted on by a different force, and for which the other cannot be substituted, is provided for sending forth this centrifugal power to act speedily, and in the culture of life, determinately, on the muscular and psychical organisms. Thus is seen the inosculation of the somatic, psychical, and spiritual lives, with their differentiate forces for action, reaction, and determinate interposition of action. § 9.

Correlations between Forces, — their constructive differentiations, I. i. 8–12, 17–31, — as they manifest their powers in and from and among and over the simple chemical substances, appear as the constant phenomena of nature. Nature teems with crystals. They shower down around us in snow, and they form in intricate figures in the solid ice. In what is called the crystalline province of nature, the solid and the liquid elements meet, and the rudiments of *forms* common to the solid and the liquid make their early appearance. In the mineralogic crystallizations the forms are all angular, and "are bounded on all sides by plain surfaces;" while in the liquids and in the first step of organic life and through its vast and multiplied ranges, curved lines and surfaces prevail. As is the ascent in this life of things, so is the predominancy of the Redactive Force in its variety of Forms. The solid crystallizations are angular, or, in other words, are the effect of forces acting in straight lines — backward and forward. The *cleavage* of crystals generally splits and separates in certain directions, disclosing polished surfaces, and showing the order of formation of successive parts. "When a crystal is broken, there is a tendency to repair it; it continues to increase in every direction, but the growth is most active on the fractured surface." So the movement forward is from an unorganized or simple condition of the elements, through crystallization, to the living, redactive forms of the vegetal and animal life. The elements of nature, before any formative processes commenced, must have been in solid, liquid, or gaseous, inorganized condition. They, then, must have moved into crystalline, vegetal, and animal forms in virtue of superinduced formative processes. The correlations for these organizing processes must be seen as necessarily prearranged for these subsequent adjustments and adjustibilities. These correlations having been provided from before and in the beginning, they orderly unfold in subordination to the intelligible forms necessary for the action of each organic part and the *form* of the specific whole which makes each thing its differentiate species. These correlations are seen pervading the entire complement of the simple chemical elements, down in their simplest forms ; and if chemistry shall detect elements simpler than these, then there the correlations for all the subsequent organizations will be found implexed and inwoven, and thence throughout to the highest organizations giving the syntactic intelligibility which binds the whole into system. They are seen in the dynamic, the plastic, the autonomic, the psychic, and they will unfold intelligibly and in system in the autopsic powers. These correlations become openly manifest in the actions and reactions of the vegetal and the animal orders.

THE VEGETABLE	THE ANIMAL
Absorbs carbonic acid from the air ;	Returns carbonic acid to the air ;
Supplies oxygen to the atmosphere ;	Withdraws oxygen from the atmosphere ;
Decomposes carbonic acid, water, and ammoniacal salts ;	Produces carbonic acid, water, and ammoniacal salts ;
Produces the organic principles of food ;	Consumes the organic principles of food ;
Endows mineral matters with the properties of life ;	Deprives organic matter of the properties of life ;
Imparts to chemical atoms the property of combustibility ;	Deprives chemical atoms of the property of combustibility.
Imparts to chemical atoms the power of nourishing the animal ;	Imparts to chemical atoms the power of nourishing the vegetable ;

Converts simple into complex compounds;	Reduces complex towards simple elements;
Is an apparatus of *de*-oxidation ;	Is an apparatus of oxidation ;
Absorbs heat and electricity ;	Produces heat and electricity.

To suggest, but not now to press conclusions which may seem too vague and not sufficiently traced in the empirical psychologies of the times as the correlations of the Forces as the Moral Powers of man, it may be seen that, as moral forces, when the explosive passions burst forth uncontrollably, the system of the Moral Forces in and for the Self is disorganized, or the forces are not in their proper correlation for acting in harmonious unisoh ; they are sharp and angular. When the attractive appetencies — love of anything — inordinately prevail, then the conduct is equally abnormal and *uncorrelated,* and again the conduct is sharp and direct and close akin to the animal instincts. But in the adjusted correlation of the explosive passions, and the appetency directed towards moral attractions, (love of order, justice, holiness,) which can only be in virtue of *their* correlation with the Intellective Power, then the normal action of the moral life is perfecting in the adjustments of these very forces of life. I. ii. 19 ; *ante,* 2, 3 ; *post,* c. vi. § 2.

12. In the provision of Efferent nerves and the determinate power of interposition by which the Self may or may not send forth, in the normal condition of the Personality, its forces of action and control, is seen the actual inosculation of the somatic, the instinctive, the psychic, and the solidaric or spiritual lives. The Self may be there in its Independency to use or not the forces inherent in its proper Self. The very delicacy

of the organizations by which the Self can act on and
through the organisms with the tremendous violence of
its muscular movements, and yet can regulate them to the
gentlest touch of movement. perceptible to others, indicates those correlations between the Self and these organizations, by which, when the organisms are out of
tune, are automacized or otherwise injured, the organism, so influenced, will react upon the Self, or at least
will prevent manifestation of the autopsic Self through
the malignly influenced organism. The maligned organism becomes automatic — self-acting. It no longer obeys
or responds to the determining Self. I. i. 32 ; ii. 11.

13. The ganglionic centres which build the somatic
body are distributed variously in the different parts of the
system, according to their respective intelligential endowments for building up the whole Form. Superimposed
upon the whole ganglionic organisms conducive to the
somatic life is a complexure of organisms in which are
inwoven the functions for action, intellections, and the
various forms of gratifications in loves ; and on these
again are placed the organization for determinate action,
determinate thought, and for bringing the love in the
Self into correlation with the Insistent truth, the Proleptic Morality, and the Divine Ideas. I. i. 35–37.

The instincts of the animal classes are represented in
the human system either as instincts or are advanced
and expanded into the open play of capacities, which in
the superimposition of a conscious Self are consciously
exercised or restrained from action. The senses are the
same, external and internal, with modifications only. Up
to the point of ideation and intuition, and, as flowing
from these, the power of *willing*, I. i. 18, for the execution of moral duties, there is no power, intellectual
organic action, or love in man, of its various kinds,

which has not its representative homologue in animal life.

14. The Self, in its human consciousness, is thus situated in the web of its vast complexure, with its filamentary subjectivities stretching out into all the objectivities of nature and life, and running as it were along these lines, catches, intuscepts their Intelligibilities. So situated, it receives, by its organisms, sensations from the outer world, and by its functionalized orgasms, acting in their appropriate organisms, sensations from its animalistic life ; as the brain-organism writes much of its activities, if not all, in one form or another on the different viscera, so it receives sensations and cognitions from and through these. I. ii. 12, 16–18. In most, if not in all, of these organisms there is much of growth and increased activity, giving increased intensity to the orgasms at work, especially in man, where this increased intensity is the result of voluntary indulgence — thus combining the action of the animalistic, the human, and the spiritual forces. And so, inversely, intellectual powers, volitional activities, and affectional intensifications are increased by cultivation, and in the mutual action, sympathies and antagonisms of life. These organisms are not, simply, completed instrumentalities converging and distributing perfect agent-forces, but they are implements improving or deteriorating under the action of the forces demonstrating in and through them. The action of the Self is modified by the external causes which affect the autonomy, and, in turn, the Self, in its social and moral culture, reacts on the autonomy, chiselling and engraving its character on the organisms, and so on the Inclusive form. § 9 ; I. i. 1 ; II. iv.

15. It is becoming manifest that the somatic life is supplied and supported, and its various movements are

continued, in virtue of assimilable substances and forces
— plasticities — taken up from the food, water, air, and
light. Withdraw these, and death to both vegetal and
animal life is the consequence. This is seen in the vegetal ; so is it in the purely animal life. §§ 7, 8. In the
latter, these elements of supply are furnished through
the stomach, the lungs, and by absorption of the capil-
lary vessels ; and the animal and the man exhausts, or
can exhaust, much of the forces thus taken up in its various instinctive gratifications and muscular action, and
the whole body changes and the devitalized elements are
thrown off entirely in every six to eight years of life.
In one drop of water there is force enough concreted to
kill forty men when prearrangedly exploded. In the organizations the brain-organisms are in like manner supplied with their requisite amount and appropriate functionalization of forces from some common source or
sources of supply. I. i. 3, 4 ; ii. 11. As long as the
animal nature is purely somatic and instinctive, the supply of the various organisms with nutrition and forces
will follow the simple currents of the autonomic laws of
growth and supply of action. If the animal is taken
and trained by man, the course, direction, and supply of
forces will be changed from this natural order, and will
be carried to the nerves and muscles brought chiefly into
use by the system of training adopted. So, when man
trains himself or is trained in the trades, professions, &c.
of life. So it is in the training in the soul-organisms,
whether in the villanies and debaucheries of life, or in
its æsthetic cultures, or in grace and goodness. The fish
in the dark waters of the Mammoth Cave, of the same
seeing species in the neighboring streams, have no eyes.
The Forces for action and manifestation are furnished
from the Plasticities, and are assimilable in virtue of the

autonomic laws assigned to each species, and, in it, to each ganglionic centre of growth or action; and nature, God, does not give anything unless for use and in use. There is no Movement without force, and these forces thus supplied must follow the laws inwoven in the autonomy, or as these are modified by geotic causes, or conscious superintendence, as in training, or by self-direction in self-training. Here it will be seen that there must be an organization, in and within, by which the Proper Self can direct, control, and countervail within certain limits the powers, the forces, thus supplied to the animal, the somatic life, and to the human orgasms, and thus manifest a higher source and quality of forces. It is here that the action and reaction between the Spirit on the one side, and the Soul and Body on the other, become manifest. The Spirit, by mortifying the desires of the Flesh and the Soul, subtracts their forces, gives them a new direction, and they die, as the fish become blind — as nations prepare for their doom, or ascend to goodness. The fact and use of moral powers can only be known and preserved by their use. If the Spirit goes with the animal man into animalistic indulgences and ingests its forces therein, they become vastly intensified; if into the human orgasms, they too are intensified; and if continued, automacy, in some form of monomania or uncontrollable fanaticism, supervenes. The presidency of the Autopsic Self, and the resistance and control of the animalistic and human orgasms, thus supplied with forces from the assimilations of nature, show the conscious supremacy of the Spirit. Yet with and from these forces it builds its new life. From the same soil, water, air, light, the tulip and the lily, so much alike in their germs, build their respective and differentiate *forms* of species.

16. The Consciousness, in its threefold correlations

of Actuation, Intellection, and Love, as herein shown, has its *locum tenens* in its organisms of receptivity and of manifestations. This threefold consciousness, in its own self-possession, is the proper Self. §§ 9, 11.

17. The Consciousness is now suggestively the unifying and harmonizing bond of the Triplicity in the Self. II. ii. 39–47. The elemental forces of the Triplicity manifest themselves in and through their own appropriate organisms and by their own appropriate Efferent nerves. The independence and reciprocal action and reaction of the psychical orgasms are separate and distinct, as is psychologically shown in that the affectional and the determinate movements are frequently in conscious opposition to each other, and that the latter frequently suppress the former; the intellective action opposes and represses both, or lends its force to their joint actuation; and it is physiologically visible in various cases, as, in hemiplegia, " where no effort of the *Will* could move the arm, it has been seen to be violently jerked under the influence of affectional agitation consequent on the sight of a friend." Carp., *H. P.*, § 622; II. v. 33.

18. Subjective Identity is the continued existence of the solidaric Self, and is accompanied by that Consciousness which takes self-cognition of its own continuity and of external phenomena and internal sensations and of its own inner operations, with more or less distinct recognition of its own perduring subsistence amid these changing variations. It is the self-cognition of the Self's abiding energizings, or rather capacity for successive energizings.

19. Objective Identity is the substratum — the substance for continuous or successive phenomena — the thing out of which they proceed. In the stabilitated elements, bodies, or forces of nature, which remain un-

changed, or unchanged for any given time, such may, for the time, be called objective identities. These objective identities may change in form and they may change in substances. In nature and life all things are undergoing change; as they change in mere form, there has been an application of forces, and as they change in substances, there have been a composition and resolution or resolution of forces; and a thing which then was, by reason of the exchange of forces, now is not. I. i. 10. It has received something or other, or has parted with something which it had, and is no longer the thing that it was. This is the " flowing and becoming" of the Greek philosophizing. There is, ontologically, no valid objective identity, except the primal creative forces; and if Immortality is accorded to man, this Self which embodies, receives it, is both a subjective and objective identity, only as the gift of God. But there is, in one sense, an objective identity for Science in this, that certain compositions and resolutions of forces as they appear in certain compounds, or compositions, do in other combinations produce other compositions and resolutions of forces, and as such are termed new substances with new phenomena. These effect-producing agents may be called objective identities, which, in a more fundamental scientific ontology, will be traceable back to simple chemical bodies and their correlating forces. Civilization and Evangelization will be seen to be but higher correlations and movements of living forces. I. i. 25, 41–43.

20. By that law of mind which is analytic, synthetic, and redactive, I. i. 38–40, by which an invariable class of phenomena are synthesized together and attributed to specific substances or to their substratum, and others to another, referring each class of phenomena to their ap-

propriate substance, and so of all (or else all processes
of reasoning are whimsical, and there is no intelligent
chasing of forces in their subtle windings, § 8) different
noumena — underlying ontologies are notionalized for
the uniform productions of their respective phenomena.
As in the fixed and successive phenomena of the material
world there are indicated various intelligential forces acting independently, as it were, or concreted and inwoven
in the substances of nature, as in the dynamic, plastic,
and autonomic manifestations, so in the psychical world
there are autopsic forces projected and manifested, each
in its own special and characteristic quality by which it
is cognized ; and these, under the same law of mind, must
be ascribed to more fundamental subsistences as their
respective sources of causations. Holding steadily,
therefore, as the thread of the labyrinth, the clue of Intelligible Forces, their subtle windings lead, not only to
these triplicate forces in the Self, but to the foundational
ontologies in the coördinate, coessential forces in the
Beginning.

21. These special and respective manifestations of
forces are ascribed, in virtue of this regulative process,
(now becoming more apparent, as the method of intuscepting them in virtue of the threefold powers of the
Self is consciously exercised,) to their diverse powers of
causation, connected into a Unity by a concordant Consciousness, and which is now, as hereinbefore, styled the
Self. This self-conscious Self, this Spirit, Πνευμα, thus
implexed with and inosculated to the organism of sensation in the body, the Σωμα, and to the organisms of the
soul, Ψυχη, and endowed in some way with the intuition
of the Insistent Truth, the ideation of the Proleptic Morality, and the Divine Ideas, projects from its conscious
unity its threefold phenomenal manifestations, in such

separate definiteness, difference, and contrasted identities
as compel the conclusion that they, each class, are ra-
tionalistically referrible to its own underlying source of
cause, and that these, in the unity of their subjectivity
in the Self and in the diversities of their manifestations,
are the spiritual personality of man.   Again, we get the
body, soul, and spirit of St. Paul. 1 Thess. v. 23 ; I. i. 1 ;
5 ; *ante*, §§ 5–16.

22. The phenomena of actuous objectifying — of going
over into life with the activities of the Self, of intellectualizing or loving, or their evolution singly in spontaneity, or their complex manifestation when *normalated*
into determinate objectivity and thus *ex*-truded into life,
and caught and detained in the memory, are subjected to
scrutiny, and thus become to the Self, itself, objects of
reflection — of contemplation in similar manner as the
phenomena derived through the sense-bearers, I. i. 2, —
such phenomena or effects being written — notated on
the different portions of the nervous system in correlate
communications with the Efferent nerves sending down
the special forces or influences. § 14 ; II. iv.   The action
of and reaction on these psychical causations upon and
by each other, and the clear and unmistakable capacity
of the conscious Self to be present in the working — the
phenomenalizing of the one and of the other, and to know
when they are in opposition and when they accord in the
Consciousness, and the physiological proof of their difference and diversity, § 14, give all the verification needed
by Science.   Thus they are resolved from their complex
action into their discrete acts of phenomenalizations, and
thus to their very sources of causations ; and thus it is seen
how the Self can set one over against the others, rotation-
ally, and from the position of each cognize the others.
In this ability to objectify each and set over the one

against the others is given the noble power of Self-Possession and of conscious Self-Direction, and, in the highest perfected condition of the Self, to regulate and control each of these psychical activities in the orgasms of its animalistic and human organisms. The eye sees everything else, but itself it cannot see ; it can see in the effects on its various viscera the reflected image of itself; and so the Self sees itself, in its triplicate elements, in their respective positions and their actual working correlations. There, these and only these are given, and these are all it will find in the moving forces of the universe, however complex they may appear as they manifest themselves in the operations of nature and life, and however comprehensive they may be in the grandeur of the highest human excellence, and however simple and sublime in the coördinate essentialities in God. I. ii. 11, 12, 19.

23. As from and by means of the sense-world and the organisms correlated thereto, sensations are *in*-truded into the Consciousness and there retained and *reproduced* as occasions prompt in Imaginates, as from the psychical organisms their indwelling and special functionalizations, as of anger, self-defence, organic intellections, and instinctive loves are manifested, so upon these, and over all, the Self exercises its autopsic control, and in determinate action *ex*-trudes, objectifies its powers into nature and in the currents of life, and all are notated, engraved in the organisms by the modifying forces of Vital Activity. *These can be reproduced by reflex acts.* It is in reënlifing, reproducing, reflectively, these effects in the organic structures, that it is clearly seen in the introspective analysis that Actuation is not Intellection, and that this is not Loving, and loving is neither the one nor the other of the two, and that all are different, and must therefore arise out of different sources of causation. They are

not only seen in their diversities, but in their different
degrees in intensities of their respective kinds; while in
the culture of life they can also be noted as susceptible
of control, regulation, improvement, and deterioration.

24. Thus there is a central Ego, Self, having a common ground of Consciousness for the reception of the
sense-phenomena of their various kinds, of the movements and impulsions from the soul-organisms, and of Intuition, and capable of self-normalation, I. i. 33, in Ideation, and which Self, in these processes or in the lower
animalistic and human impulsions to conduct, ever and
always works out into manifold demonstrations of its
special, complex organization.' From the rudimentary
germs, the organisms increase in capacity, and expand for
action and intensification as is the culture of life, and as
the forces of life are turned in this direction or that, upon
these organic functions or upon those. Thus are the
final judgments of life made up; " therefore thou art in-
excusable, O man, whosoever thou art, that judgest : for
wherein thou judgest another, thou condemnest thyself,"
for if an animal, as tiger, monkey, or wolf, could be supposed to judge another, it could only judge *from* its own
nature ; if man judges another from the dictates of his
animalistic nature, the judgment will partake of the elements in the man from which the judgment proceeded ;
if he judges from his natural elements of malevolence,
envy, covetousness, pride, self-love, so these elements
will be involved in the judgment, and God can only judge
him by these elements thus working in life and judging
in the man ; but if the judgment is one of charity, meekness, and love, then the divine judgment finds and accords
these in His judgments of mercy. ".With what judgment thou judgest, it shall be judged to thee again."
Thus, at every turn of the unfolding manifestations, the

diversities, complexities, and their unity in and around the Self, appear.

25. The system of fundamental correlations in and around the Self, thus more fully set forth, unfolds the following propositions to be established and further elucidated.

a. That, beside the sense-phenomena communicated by the external and internal Senses and the impulsions of the Soul-orgasms, the only elementary *facts* submitted to and within the subjective observation of the Self are its own subjective manifestations. 1. Of doing, objectifying — creating — in the sense of the Self creating and setting over in nature and life its own facts — *facta* — deeds, in some of its many modes of Actualization. 2. Of intellectualizing, in some one of the many modes of exercising its Intellectivity. 3. Of loving, in some of its many modes of gratification — even in the appetencies of its deepest malignities. I. i. 21. All these subjective manifestations, whether of meditation, contemplation, loves of the vicious or sinful gratifications, or actualized manifestations — the creative facts of the Self in which all the threefold powers of the Self conjoin, are resolvable into one or the other of the foregoing trinal elements, or into their complex and conjoint action as causative efficiencies.

b. That these functional or radical phenomena will, in these trine aspects, be found to be radically diverse; that, as in the organs of sense there is but one eye and three different colors in nature — red, blue, and yellow, and but *one* light yet with many shades or intensities of color, so in intellectualizings there are many forms, but one Intellectivity; as there are many modes and forms of acting and but one Actuous Power, so is there but one Love lying at the base of the many gratifications, —

and all these are complexed when the Self goes over into nature and life for subjective gratification. It is in some form of Love that Motives are found, and Sentiments are moulded by the Intellectivity as the attractive end in the Self for its action.

c. These hypostatic elements are personative of three unmade, coequal, coeternal, and consentaneous causative essences in the Creator — the Trine in their coördinate Unity. These elements of Personality in man are symbolic and representative. As the Self cannot conceive or ideate anything as made or fashioned without objectiv-facient force to stabilitate and set it over in nature and actuate it; nor this as done intelligently in its form of existence and actuation and in its various correlations without an Intellectivity from on the other side of nature and life, nor as intelligible to intelligences on *this side* without this Intellectivity in both ; nor these as moving forward in their Creative Intelligence without some motive sentiment — some end of gratification — some love in the doing and accomplishment of the creating intelligence, from which it lights up and enlifes the Intelligibilities thus created ; and as it finds these in itself, so can it find no other essentialities in the Personality of the Godhead, and these it must find as Causative Efficiencies, for without positive efficiencies as forces they are causeless and cannot produce phenomena — facts. I. i. 8–10, 25. This Self is the image of the Divine Self. When "born again" by that process of life which gives the Self, in the supremacy of its solidaric self-consciousness, control of the animalistic and human orgasms, and subjugates their spontaneities as such, and enlifes its conduct with the highest form of Love, disenveloped from these orgasmic impulses, it attains at each step a higher life. As the human life, influenced by human motives and sentiments,

is higher than the animalistic life, and yet is only human,
so is the Spiritual life higher than the human, and the
ascent is an approach nearer to the similitude, the likeness of God. It is thus renewed, restored to the pre-
eminence and mastery of its solidaric, its spiritual life, by
disenvelopment from the control and orgasmic influences
of its lower *forms* of life. This is and can be attained
only in the *love* of a more harmonious order of life — a
higher *right*eousness of existence ; and Love, in the progress of the ages and the disciplines of sorrow and sympathies, and in the harmonizing effluences of this Love,
as it rises with healing on its wings from the wrecks
and ruins of kingdoms, empires, and republics, moulds
societies into higher organizations than the mere secular
institutions, and *re*-forms man into the likeness of his
Creator, God. The causes, the potent efficiencies
which do operate now, and which have operated in the
centuries heretofore on the races of man, will continue
to operate upon and effect their autonomies and mould
them and make them sanctified or unsanctified instrumentalities of their perduring solidarities. Governments but restrain the excesses of these animalistic and
human orgasms ; but the love of Order, Justice, Righteousness, which will be seen as synonyms, controls, subjugates, and sanctifies them into a higher life ; — and
" having spoiled Principalities and powers, he will make
a show of them openly, triumphing over them in it."

d. This Actuocity, this power of actuation into the
deeds of life, this Intellectivity choosing between this
love and that, and giving forms of action or speech, and
selecting time, place, and means of action, and this Love,
broken up and functionalized into many forms of gratification in the diversified and complex orgasms in the
autonomies of animal and man, are the only elements

manifested in the lives of individuals and in their tribal and historical movements. This underlying spiritual Triplicity contains all the movement-elements of the Proper Self manifested in history, whether appearing in superstitions, philosophies, or forms of religion, or regulating or improvising individual conduct, or in the theocratic dispensations, or in any of the forms of governments of men. At each step of improvement, the advance is but the evolution by self-normalating control, and the reaching up to *this* higher life.

e. Sense finds man an animal, and keeps him so; philosophy ends only in æsthetic culture; and morality only in self-culture and self-control; but the complete and final elevation of the Self, of all selves, is in a love of spiritual kinhood and harmonizing sympathy for a higher — the higher spiritual life, in which the coequal and consentaneous harmonies of these Triplicate Forces unfold, are unfolded into the exaltation of the Moral Will. I. i. 17, 18; ii. 19. This is the philosophy of Grace. I. vi. 44–49.

26. Herein the term Self means that solidaric Identity of all men, subsisting in the Consciousness, however observed, and not yet unfolded into self-consciousness, and which, when this is accomplished, is seen as above the animal orgasms, yet receiving and noting their impulsions and demands to gratifications, as also above those soul-orgasms which impel it to human conduct for the gratifications of mere human life, and yet, in the new and higher life, controlling, subordinating, and using these for the attainment of higher ends, as cunning in the animal is but cunning in the man, and cunning in the bad man, who uses it for fraud and chicane, is the same cunning for the sanctified arrangement of means to lead others, and in himself to attain purified ends of life. This

Self, attaining this higher reach of life, presides over the forces of nature and life in its prescribed circle, and moulds them to uses in the discharge of duties, as in its lower life it uses them in the violations of duties — of responsibilities. In its superintendency it normalates its spontaneities of Passions and Affections (I. i. 17, 18 ; ii. 11–15) into Reflective Consciousness in the presence of its Intuitions and Ideations, I. i. 35–37, and reaches the divine harmony of life. The judgment of the Just. §§ 9, 24.

27. Being will represent God in his trine coessentialities of Creative Power, Intellective Wisdom, and Coefficient Love, and these essences as coessential forces coördinating each other, and their conjoint coaction as indicating the definite Personality of God in the effluence of his determinate Will, (I. i. 17, 18,) coming over into the objectivity of nature and into the providences of life.

28. Rational Philosophy, taking its departure from the standpoint of the single element, the Intellectivity, can never reach any other Science than that of the Formal Logic, and can never give any other philosophy of nature and life than that which may be made by putting together intellectual fragments of the universe ; the life, the action, the passions, the affections, must, will, of strict *logical* necessity, be wanting. The Moral Will is relegated. It ever has been and is incapable of reaching or demonstrating the coessentialities in the Personality of God, and hence, as a Rationalism, has in all systems, in all ages of philosophizing, ended in one or the other systems of fatalism, the Materialistic or Intellectual, or in necessitated contradictories, or, for the want of a true method, in unlegitimated mysticisms. The Christian religion, through the veil of its many formal sects, conflicting in details but not differing essentially

in fundamentals, having constantly adumbrated the divine image and likeness as being in man at his creation, and having preserved a wonderful memory of past traditions, or inspired by as wonderful presentiment of a great and glorious future, it is now possible, out of the elements furnished by religion, philosophy, and history, to reach what is fundamental, and redact the whole into a demonstration of what are the coördinate coessentialities in the Personality of God. If these are attained in such manner as to reach Creative Causes, then, however subtle the windings may be of these causations, underlying and forming and actuating the various coverings of nature and life, the Secondary Causes (I. i. 11) at work will be seen as only having vitalities and forces and forms from the primordial Creative Causes. The Formal Logic will give way to the genial and inspiring, but inexorable processes of the Moral Logic.

29. The Formal Logic, reduced to its simplest and yet complete terms, the Pure Logic, can only be a logic of Forms for Quantities, whether in Extension, Weight, or Number. In its *pure form* it is the process of adding — multiplying; or subtracting — dividing, *supposable* extension or weight. For these again resolve into but two processes: putting together — synthesizing; and separating — dirempting — analyzing. Multiplying is but a form of adding, and dividing is but a form of subtracting ; and all other mathematical processes are but *intellective forms* for applying these elementary processes. In their *concrete logic*, measurement is but adding surface to surface, or subtracting them, or actually both in one operation ; weighing is but adding or subtracting quantity, in another form, as quantity in weight ; number is but abstract aliquot parts of some positive or some given or supposable whole, even when the Self shall attempt to conceive

infinity. When the ultimate impenetrability of matter is conceived, quantity, as representing extension and weight, will be seen as precise equivalents of the constituting, the creative forces which have entered into the construction of matter. If matter is a composition of forces, the extension and density and consequent weight of matter will be precisely as these forces are concreted — compressed in matter. I. i. 8–10, 6, 12, 25, 29 ; ii. 6, 8, 9 ; iii. 17, 18 ; v. 15. This will give extension, weight, and number as precise equivalents — let these resolve inscrutably yet into the Moral Forces. This Pure Logic is but one of Form to be filled with actual or supposable quantities, Forces. But when the Self is turned from these Pure Forms to the actual material content, represented by them in the objects of nature and life, the elements of use for an End, a causative wherefore in the End, inducing the creation of these things, in the love of the Creator, or in the gratifications of the created or in both, become essential and necessary to the thought of their creation. It is a necessity of Moral Thought. There is no Sense of Moral Obligation in the Formal Logic, either in its subjective processes or objective Forms ; but the moment the Moral Logic enters the mind, the inquiry comes, wherefore ? to what end ? what causative power induces to the production, continuation, and to the End as a *reciprocating* cause ? I. i. 15. When the Moral Logic, in its great fulness, enters the mind, it becomes genial, inspiring, and inexorable in its preservation and demonstration of the Moral Life — the Order, the Justice, the Righteousness of God. God in his Logic starts from his Omniscience, and never errs ; Man starts in his analytic ignorance, and is but seldom right. God knows — Man learns. God alone can execute his order, justice, Righteousness ; he alone can claim vengeance — *his* vengeance which is satisfied with returning love suffusing and actuating the life, for he never errs, and this is all he can ask or does ask of man. I. ii. 19. But man, standing in his place in the on-going prolepsis, and imperfect in his knowledge, confused, perplexed, bewildered, violently controlled by the passions and affections, or some predominating one inwoven in his animalistic nature, and from which he is to be disenveloped, finds so frequently only a system of narrow and intense animalistic reason, unless in higher conditionings, for the gratifications of his human propensities, or. the selfish and therefore divinely illogical justifications of his human ambitions, politics, pride, covetousness, and for the gratification of his ideational monomanias or fanaticisms — of their multipled secular. and ecclesiastical forms; and he only gets the genial and unfolding and, also, but in a holy sense, inexorable Moral Logic from the Martyrs of Love, (Paul at Ephesus among wild beasts, Peter on the cross,) sustained by *the* Love that disowns all native kinhood with Fraud or Force, and in earnest but simple meekness controls these animalistic and human orgasmic forces, and actuates the spiritual life of Moral Love into the currents of human life. As man aspires in this simplicity, he reaches up to the Primal Moral Logic. He will then, and not till then, understand the Method of Christ in his interview with Nicodemus; for there must be a birth above the animal birth, above the human birth — a birth of the spiritual life, by which it can subordinate those and unfold into higher love. That which is born of the flesh is flesh, and that which is born of the Spirit is spirit. § 24.

30. To show the mazes in which the mind may travel in the pursuit of the Initiate Identities, and following the lines of special investigations as flowing from partial or

truncated or broken and complicate and, therefore, unanalyzed elements of activity, and thus appreciate the
necessity of gathering the colligating bands of the entire
system into one concordant bond of union, it is proper
to state, in a brief and authentic manner, the ends and the
processes involved in all the systems of philosophizing
heretofore pursued, and these as giving, inferentially,
their methods of investigation.

It has been said by accredited authority, (Mansel's
Pro. Logica, ch. ix.,) " That in the investigation of
mind as well as matter, phenomena are alone the legitimate objects of Science; the substance and essential nature of both being beyond the reach of human faculties.
Whereas Metaphysics has, from the earliest days, been
distinguished as the science of Being, as Being in opposition to " (rather as lying under) " all inquiries into
the phenomena exhibited by this or that class of objects.
How far such a problem is capable of solution is another
question; but the mere propounding of it implies an
object totally distinct from that of an inquiry into the
faculties and laws of the human mind."

" The *object* of the older Metaphysics has been distinguished in all ages as the One and the Real " — the
το παν — " in opposition to the Many and the apparent "
— το οντα, the phenomenal — the flowing and becoming. " Matter, for example, as perceived by the senses,
is a combination of distinct and heterogeneous qualities,
discernible, some by sight, some by smell, some by touch,
some by hearing; " and he might have added, by taste and
by their pathogenetic effects on the system in producing
diseases and curing diseases, and acting specifically on the
various organs. " What is the thing itself, the subject
and owner, each, of these several qualities, and yet not
identical with any one of them ? What is it by virtue

of which these several attributes constitute and belong
to one and the same thing," and what are those correlations by which they act and react, unite and dissolve,
produce disease and cure disease, and inflame or soothe
the organisms? "Mind, in like manner, presents to
Consciousness so many distinct states and operations
and feelings. What is the nature of that one mind of
which all these are so many modifications? The inquiry
may be carried higher still. Can we attain to any single
conception of Being in general, to which both Mind and
Matter are subordinate, and from which the essence of
each may be deduced?"

"Ontology, or Metaphysics proper, as thus explained,
may be treated in two different *methods* [?] according
as its exponent is a believer in το ον, or in το οντα, in one
or in many fundamental *principles* of things. In the
former, all objects whatever are regarded as phenomenal
modifications of the One (το ον) and same substance,
or as self-determined effects of one and the same cause.
The necessary result of this method is to reduce all
metaphysical philosophy to a Rational Theology, the one
Substance or Cause being identified with the Absolute
or Deity. According to the latter method, which pro-
fesses to treat of different classes of Beings independently,
Metaphysics will contain three coördinate branches of
inquiry : Rational Cosmology, Rational Psychology, and
Rational Theology. The first aims at a knowledge of
the real essence, as distinguished from the phenomena
of the world" (and therefore as the inquiry after the
το παν). "The second discusses the nature and origin
— as distinguished from the faculties and affections of
the human soul and of other finite spirits ; the third
aspires to comprehend God himself, as cognizable *a priori*,
in his essential nature, apart from the indirect and rela-

9

tive indications furnished by his works, as in Natural Theology, or by his word, as in Revealed Religion. These three objects of metaphysical inquiry — God, the World, the Mind . . . ."

"The former of these methods, aiming at a knowledge of the real essence, is the bolder and the more consequent; and, moreover, the only one which can be consistently followed by those who believe in the possibility of a philosophy of the Absolute. For a plurality of real objects being once admitted as the highest reach attainable by human faculties, these must necessarily be regarded as related to and limited by each other. Accordingly, this method has been followed by the hardiest and most consistent reasoners on metaphysical questions ;

• by Spinoza, under the older form of speculation ; and by Hegel, after the Kantian revolution. But thus treated, metaphysical speculation necessarily leads to Pantheism ; and Pantheism, at this elevation, is, for all religious purposes, equivalent to Atheism. The method is thus condemned by its results; and the condemnation will not be retracted upon a psychological examination of its principles. Its fundamental conception is not thought, but its negation. The Thought, which is identified with Being in general, is not my thought nor any form of consciousness which I can personally realize. My whole consciousness is subject to the conditions of limitations and cor-relation of subject and object. A system which commences by denying this cor-relation starts with an assumption concerning the possible character of an intelligence other than human, and consequently incapable of verification by any human being."

"The second method of metaphysical inquiry (Rational Psychology) is less presumptuous, though perhaps also less consistent. It starts with the assumption of a plurality of Beings, thus virtually abandoning the philosophy of the Absolute. This plurality is virtually manifest in the contrast between the subject and the object of Consciousness, between the Self and the Not-Self, as related to and limiting each other. But the consciousness of the *cor-relative* and the limited suggests by contrast [?] the idea of the absolute and unlimited ; and thus gives rise to three distinct branches of metaphysical speculation: the *Ego* being identified with the substance of the human soul — [Spirit] — as distinguished from its phenomenal modes; the *non-ego* being identified with the reality which underlies the phenomena of the sensible world ; and the absolute or unconditioned with the Deity."

" The last of these three branches, commonly known as Rational Theology, endeavors, from the *conception* of God as an absolutely perfect Being, to deduce the necessary attributes of the Divine Nature. . . . . It was the opinion of Kant, as well as of Reid and Stewart, that the subject of mental as well as of bodily attributes is not an immediate object of consciousness ; in other words, that in mind, as in body, Substance and Unity are not *presented* but *represented*. Those who accept this doctrine are only consistent in regarding metaphysical inquiry in all its branches as a delusion."

31. This summary gives the *résumé* of the history of twenty-five centuries of speculation. The same conclusions have been substantially affirmed by Schwegler in his " History of Philosophy," and expressly by Lewes in his " Biographical History of Philosophy." It is the common confession of learned men. Philosophers have been pursuing Substance, but grasped only phenomenal shadows. Of these various hypotheses or methods, as they are, without exactness, called, it may be briefly stated that :

a. Rational Theology begins with a transcendental
assumption of its so-called, but unanalyzed, attributes
of Deity, in which the theologians, not analyzing their
own complex mental phenomena into the simple, discrete,
elementary forces, out of which the spiritual phenomena
arise, so as thereby to be able to affirm that they have
arrived at the fundamental causations out of which these
respective phenomena are manifested in and form themselves and unite in complex correlations, have not been
able to reach the Essential Causations concerned in the
work of creation.   They anthropomorphize Deity by
ascribing to Him the complex and concreted elements inwoven in and functionalized — adjusted, and correlated
for the limited agency and responsibility of man in a
theatre of constantly changing vicissitudes of invincible
causes and effects, and these depending to a great degree,
so far as this limited agency and responsibility are involved and many of the economies of nature are concerned, on philosophical contingencies, producing, ever
and forever, new combinations in nature, and new and
ever-recurring opportunities and necessities for the exercise of responsibilities by man for the acquisition of moral
Rights and the discharge of increased duties — responsibilities, thus elevating him in the knowledge and the
exercise of his whole nature as correlated to the cosmos,
and as capable of growth and expansion from zero to
archangelic capacities, — this is the progress of individuals of all races from their inwoven orgasms to the complemental disenvelopment of the solidaric spirituality, —
the movement of tribes and races from the nomads of
Asia and the savage denizens of Africa and elsewhere, in
the education of the accumulated centuries in advancing
forms of social and evangelic perfectibility.   Rational
Theology, following its unphilosophical and unintuscep-

tible direction, will run a continual round of unfruitful labors, clutching at shadows, and holding nothing substantial in its faith and practice, save the natural longings — the appetencies of the heart for something holier and better, if it does not lose these in the disappointments in its mere abstract processes and in its practical failures for the reconstruction of society. Man sows to the animalistic, and reaps the animal; he sows to the human, and only gathers the bitter fruits of covetousness, pride, ambition, and the crops of human follies. The ruins of the system are all around us everywhere in the want of that consociating, educative, and positively coördinating Love which soothes and harmonizes, instead of exacerbating, by fanaticisms, these animalistic and human indulgences, — the passions of secularization. In this aspect of its philosophizing it is but a theory of empty and *forceless* generalizations, giving no Efficiencies for a Beginning, and a perduring and a proleptical *working* of nature, life, and providence. It does not give to man a *method* of inweaving into his life those spiritual forces which, like the action of all known forces, shall alter, change, and reconstruct the tissues and orgasmic powers of his Body and his Soul, and unfold the Spirit in its sublime simplicity. I. i. 1, 10, 18, 34, 42; ii. 5, 6, 10, 11; iii. 1, 3–15.

Rational Psychology, heretofore dealing with the το οντα, the many-faced phenomena, which are incapable of being referred to one single underlying identity for such variety and discriminate classes of phenomenalizations, and incapable of finding, in its partial and fragmentary method or process, diverse powers with a positive force of causation for coördinating and harmonizing consentaneousness for the diversified forces and conflicting strifes in nature as well as in life, leaves upon the surface of history and society not its wrecks and its ruins, but its disjointed and formless parts, incapable of a syntactic unity of correlated adjustibilities and working efficiencies, in the absence of such coördinating Powers. §§ 21–23.

Rational Cosmology, in violation of all the laws of thinking, beginning in its philosophy of the Absolute, its one Identity, its eternal Homogeneity, and positing its absolute Unity, its Identity, its One — an unthinkable source of cause, without difference of subsistences and therefore without diversities for causation, must begin and end in *nihil* — in nothing but its absolute Oneness, incapable of Causation, as without creating energies or diversities of ontologic creative forces. Or if this Rational Cosmology starts with the Many — the materialistic plurality — το οντα, in like manner it has no method or process for finding a power or powers of coördinating and harmonizing control to furnish forth the concordant adjustibilities and the harmonizing strifes or subordinated contra-pellences of nature and life, nor the evolution of moral, intellectual, and affectional action in their discrete differences and in their complex action as Determinate Moral Will.

32. Disenvelop the Forces of nature and life, as they run back in recondite but not wholly hidden filamentations, from their clear sunlight in the autopsic Self, down through their connected and inwoven forces in the human orgasms, the animalistic and the animal instincts, in the wisely weaving forces of the animal and the vegetal autonomies, in the intelligible correlations of the plasticities, in the wonderful dynamics which poised and hung the planets on nothing in their prescribed places with mass and weight and tangential projectility, or before they had mass and weight — to the Autopsic Mover of all Forces — and, then, from Him back to the autopsic Self,

consciously cognizing these forces and appropriating them to his own use in creative structures of meditation or deeds, and, at both ends, they resolve into the power to do — to objectify forth from the Self, the intellective force to do and objectify wisely and well, and a Love which coördinates their movements. And these, in their respective spheres, use and rule the intermediate Forces.

BOOK FIRST.

THE ARCHITECT.

—◆—

CHAPTER FOURTH.

METHOD. — SYSTEM. — FATE. — PHILOSOPHY. — GOD.

1. ALL true Processes for arriving at a System of the Universe, then, are based on the Method of Intusceptive Analysis. And this Method must rule all processes in every inquiry into the constitution of nature, its general action, or into the *being* and nature of God. In matters depending on the conjunction or combination of Philosophical Contingencies, I. i. 13, the most probable element or elements on which the conjunctions or combinations may take place must be seized, on which the *opinion*, the hypothesis of possibility or probability of eventual action may take place or has taken place, and this so as to exclude contradictories in the theory or the details : in Transcendentalism, it is the systematic correlation of Ideas which will harmonize the actual facts in the phenomena to appear or which have appeared ; in a Proleptic History of the solidaric humanity it is the thread of catenation, the syntax of correlations for the Prolepsis which binds the whole into a grand and ordering and orderly movement of life ; while in Fundamental Philos-

ophy it is the cognition and intusception of all Forces
in their root-forces — the causative essences, which lie
at the foundations of all movement at the base of the
elemental system and govern its growth, of all that
which is of growth, and moulds, within certain limitary
lines, all that which is of autopsic normalation. In this
their highest synthesis, all these are harmonious and
consistent, and rule within the limits of their assigned
Prolepsis. The comprehension of these phenomenalizing forces — the actual grasping and verification of the
fundamental movements of nature and life, or to reach
to the Initiate Causations still beyond, the grasping of
the Final Ontologies in their root-forces, can only be
attained and comprehended by that analysis which takes
off concrete after concrete, covering after covering, and
beholds in their serene and clear essences the Forces as
they well up from the eternal deep into the creations and
living forces of nature and life.

2. When the Method of a fundamental philosophizing
is the ideation of a cohering, as a simple ongoing or
catenation in formal logical cause and effect, it will be
purely ideological or rationalistic; that is, it will arise
out of the purely intellective processes, and will give,
when turned exclusively to the side of nature in its
material aspect, the unbroken chain of physical cause
and effect, *yet without any intelligible working contents to
causes and effects* — without the valid conception of the
intelligible forces ; when confined exclusively to the pure
operations of the Intellectivity, it will give Metaphysical Necessity — Intellectual Fatalism ; and when the
. Affectional character prevails in a strong love of goodness and purity, which the depraved world in its activities
does not realize, and the Intellectivity cannot sufficiently
fathom, in logical processes, the mystery of guilt and

goodness, and legitimate them in any consistent theory
of life or the universe, it will end in Mysticism.

3. The conscious interpenetration — intusception of
and into material phenomena gives their law-forces —
forces which work blindly wise, in given forms, and
producing certain results, which, when redactively formulated in the Intellectivity and enounced in words, are
called the laws of matter. I. i. 24–31, Here the Laws
of matter are reached *a priori*, but by ascending through
the analysis of the facts as they flow from the forces,
and binding the facts, in the synthesis of their respective
sources, the Self ascends to the transcendental standpoint and obtains the Divine Ideas which gave the forms,
and thus the substantive and constitutive forces of the
things which produce facts are necessarily ideated. Law,
in this sense, is the ideation, the actual intusception of
powers — forces above and preëxisting the creations — the
facta, from those forces. It is the antetypal ideas and
the forces for their actualization for the forms, organisms,
functions, offices, and capacities of the subsequent creations. It is a transcendental Intellectivity giving to
matter the special and particular forms, and transcendental forces — powers giving substantive constitutions,
in primordial and secondary and intermediary differentiations, by virtue of which they maintain their orders, &c.,
and produce their specific characteristics and their diverse
actions ; and this whether the specific symbol is a pro-
notozoon, a man, a planet, or an angel. I. i. 8–10 ; iii. 1–4.

4. But if the formal statement of Law is of a power
or powers in matter, *in et per se*, in and of itself, it is not
a Law, a transcendental idea preceding the formation and
ruling its action and organic movements, but is only a
purely subjective formula, gathered and generalized *a
posteriori* from the uniform condition and action of mat-

ter in itself and in its developing organic movements. It is but a knowledge of how matter acts by *its* forces, and does not at all, in correct modes of thinking or expression, give the ideation — the valid conception of powers above matter producing it and imposing on it regulative *forces* which it must obey. It is but an *a posteriori* knowledge obtained in the Intellectivity. The distinction is valid and important. Thus : " Of the nature of Gravity we know nothing. We give the name of Law to the *effects* which it produces." — Smithson, *In. Rep.* 1858, p. 105. And so must all writers say, until they can see Powers above Matter, making it and functionalizing in organic and orgasmic forces. In strict thought, it excludes precedent, transcendental ideas of formation or creation, and so is only a subjective formula from on *this side* of nature and life, and is not a law — idea of forces *above* matter — from on the other, the creative side. On the ideation of a creation, the fore-plan must precede the executing force to bring forth — actualize the fore-plan : herein, the fore-plan is of the Intellectivity, and the working efficiencies are of these positive forces as they are differentiately functionalized and concreted in the symbols and forces of the cosmos.

5. The ideation for the movement of nature and life must be of a developing Essence or Essences, or of a Creative Personality — normalating the laws and the forces of his creation. This affirmation will present the inquiries : *a.* A mere Power or Forces to transform or in transforming ; *b.* A power and a self-impulse to produce — to emanate ; *c.* A power — force, and a spontaneous Intellectivity to produce, transform, correlate, and continue ; *d.* A power — force to actuate, and an Intellective Force, determinate in means and ends, and in time and place, in unison with some coördinating impetus, *appetitus*

animi, coessential love as a power, as the source and the causative end — subject-motive of creating — Love in creating for the wherefore of creating, — the why, in some gratification, *which for* he should create. I. i. 15. These are the root-ideas of all possible theories of nature and life. And they all require root-forces.

6. In the first instance it is Chance, in its lowest terms, as will appear in the subsequent considerations. I. i. 14. In the second, it is a spontaneous germ-force, from which the whole systems of things of the cosmos sprang, grew, ramified, and racemated from seed to stock, to branch, to the clusters of independent systems of existences, by orderly, coherent, and necessitated germ-forces, but without a proleptic and prearranged order of production ; a geometrization of the universe without a determinate geometry ; physical forces producing effects without the effects forecasted in intelligible series, or intelligent in the adjustments and correlations of the successive appearances of the differentiate forces of nature and life which are to act and wait on each other in vast systems of adjustibilities, and which are not less significant in their antagonisms than in their harmonies ; and causes called Moral, without the vital principle — obligation of right and wrong ; a power and an impulse to action, blindly producing forms and movements of intelligibility without intelligence to produce and without intelligence to comprehend them ; a producing development, unfolding into conscious loves without a conscious love for their production or a conscious love for their attainment and reciprocal recognition ; an infinite power of actuation and endless ongoing in unbroken development without definite means, and not, as in man, autopsically applied in details, I. i. 33, 31, and therefore without intelligent ends or final cause ; a systematic correlation without a cor-

related system; an intellectivity without an intellect. I.
i. 15, 36, 37, 41–43. These hypotheses are of chance-
medleys or of germ-developments in the orders of nature,
and of unpredeterminate and therefore of unintelligent
spontaneities in the orders of mind. They are in the
former chance-begotten; in the latter, the ongoing of an
elemental germ, or, as in some of the ancient cosmog-
onies, "an egg of night," or, as in a modern system, it
is the "primitive egg with its germinating vesicle and
germinative dot, indicative of the universal origin of all
animals." As the chance-medley of atoms, it marshalled
into harmonious systems of intellectual correlations contrapellences and dependences; or, as the great germ of
the cosmos uncoiled from its rudimentary state, in the
silent and motionless eternity, the orders of nature were
developed without a ruling and controlling Intellectivity,
in their proportionate and magnificent classes of differentiation, in a chain of invincible necessity. The logical
mind cannot see how there can be beginnings or breaks
and new beginnings in such a succession, even with the
datum of the germ-beginning, and which, thus more conclusively, excludes all new successionings. There can
be no classes of differentiation, and the constant relation
of fruit to branch, of branch to trunk, of trunk to root,
and of root to germ, must pervade the entire series of
the outgrowth, without intelligible beginning of the whole
or beginnings of the differentiated classes and orders.
The earth, teeming with the marshalling of atoms, in this
impulse of emanation or development, in historic ages
might, most surely, in the revolutions of the geologic
periods, in the multiplicity of the differentiated orders
and classes preserved in the stone leaves of nature's
record, in their emanative and primitive embryologies,
would, present facts of development imperfectly made,

and of crudities in their transitional state emerging from
the self-impregnate earth or passing from the inferior into
the superior forms, as the distinct fact of outgrowth, somewhat after the fashion of the fantasy of the poet, when but

> "half appeared
> The tawny lion, pausing to get free
> His hinder parts."

7. In the third instance, Infinite Power and Spontaneous Intellectivity can only give a developing contradistinguished from a normalated succession of cause
and effect. The Formal Logic and the Moral Logic must
both be wanting, — the former from the want of forecast
to reason to its end of action, and the latter for the want of
a moral end of action. I. i. 33 ; iii. 28–31. Mere Power is
blind ; and mere spontaneous intellectivity is not a process
foreseen and prearranged, and this, in its present form, is
spontaneously intellectual Fatalism, and is fitly embraced
in the various forms of Pantheism and some of the systems
of German Rationalism, in which God is represented as
arriving, through nature and life, at self-consciousness.

8. This Fatalism is distinguishable in formal hypothesis, and, as such, is to be distinguished from that view
of fatalism intimately connected with, if not inseparable
from, all systems of strict Rationalism, in all of which
the Self, in its pure logical processes, ratiocinates, by *a
posteriori* processes immethodically applied *a priori*, the
fore-plan of all movement in nature, life, and history,
and, in its rigid intellective *post*-ordinations, leaves no
room for the divine economies of philosophical contingencies, I. i. 13 ; ii. 3 ; *ante,* §§ 3, 4, of means for ends in the
introduction of new forces of differentiation, or direction
of forces, or for moral movements in the action of the passions and affections. This latter is that philosophizing
of the Greek stoics and others which makes God the

servant of the invincible necessity of this purely abstract and uncomplemental fore-plan — this predeterminate foreordination, borrowed *a posteriori* and applied *a priori*. He is the δουλος Θεος αναγκης — the servant of necessity. As Seneca said, "One and the same chain of necessity binds God and man; the same irreversible and unalterable course carries on divine and human things. The very maker and governor of all things, *that writ the fates, follows them*. He did but once command, he always obeys."

9. When the ideation shall be of Being with a power, force, in a determinate volition, to create — to objectify creation forth from himself in objective and stabilitated immanence and actualizing phenomenalization, — in the very intusception of such a volition, a beginning in creation, a time when the particular determinate volition was executed and manifested in symbolic forms, in their assigned places and orders, there must be a complement of forces. In such ideation there must be succession, and there may be breaks and pauses in the successions for the successive existences in their appropriate orders and successions, and for their longer or shorter duration; there may be varieties of independent organizations suited to their theatres of existence; and independent organizations with their various correlations, of necessity, imply determinate actuations at the times and places of intervention under intellective guidance foreseen and forecasted for a motive-end in love — or hate! for which the physical chain of cause and effect shall be broken, and new orders of movement introduced tending to this motive, causative end. These organizations may be differentiated, not only in degree, but kind; and so differentiated that they cannot, by any process of development or generation, be resolvable into others of a differentiate kind; and yet others may be so assimilated in parts of their organisms,

that by cross-breeding and affiliative correlations they may produce new kinds. Many of the kinds may be dependent on others in the order of their appearance, not as a rigid metaphysical succession, for the fact of correlated differentiations destroys this metaphysical as well as the spontaneous intellective successions, but as constituting a proper preparative state and supply of food, out of many such which might have been adopted, I. i. 37, all of which shall give intelligible evidence of being the product of some kind of special creative forces with designate adjustments and adjustibilities, and from the same common designer of intellectual forms, and from the same motive-sentiment inducing to the work of creation and appearing in the harmonies of their coincident correlations and oppositions — their attractions and contrapellences in their classes and orders. There may be, here, beginnings and endings of series, breaking the metaphysical chain which, so frequently, has bound noble minds in the treadmill of an endless and unalterable necessity; but this can be done only by the ideation of determinate volitions, and this only under the coördinations of the coequal and consentaneous objectiv-faciency, intellective force, and a motive-end in the gratification of the force which tends to produce, and in Deity is the exercise of motive, causative love.

10. From such facts, ascertained to exist in the geology of nature, and which have their approximative parallels in the rise and fall of races, I. ii. 12, 13, and from the tribal and historical manifestations of psychical phenomena in man, and from the accumulation of analyses lifting up to observation the triplicate powers of the human solidarity in their deobscuration from the primitive condition of the races, the analytic process may now pass to the synthesis of determinate acts of volition in the Self, and

may further, with reverential boldness, ascend and find determinate acts of volition, I. i. 17, 18, introducing and inworking the coördinate forces of the Creator into the successions, in their times and places, in the geological changes in the earth and in the historical movements of the races of men. Determinate acts of volition given synthetically *a priori* in a work of creation, then, intusceptively to the triplicate Self, such a world as this with its varied phenomenality and other worlds in endless variety of organization and function, are potentially and actually conceivable — become a propriety of thought. I. i. 9 ; iii. 28–31. The created thing must indicate the Creator ; it must be as he made it ; and he can make it only in virtue of the causative forces by which he can make, and, if they are different, of the essences or forces from which he must make. In such a creating Being there must be objectifying force, the precise equivalent of power, force in potency and to be brought into actualization to create ; there must be intellectivity as an active and modifying coördination of this force to supply and impress forms and give functions with differentiate characteristics, and not merely as the pure impersonal Insistent Truth, I. i. 35, but in full possession of it and with transcendental intellectivity to furnish and arrange, from the Divine Ideas, the vast and complicate systems of economies in the proleptic movement — the beginnings and endings of periods when new orders shall commence and old ones shall cease ; the formation of the original elements of construction for the bases of matters and the supervening vegetal and animal organizations through the long unfolding ages ; the aggregating of the primordial elements into the dynamic systems, followed by the ever-changing plasticities, the stabilitated autonomies, and the self-conscious autopsies ; and, in this process of progress,

10

the breaking up of those primordial elements which had
seethed and settled into an early plutonic condition, and
moulding their more "flowing and becoming" elements
into. vegetal and animal kingdoms, orders, classes, and
species, with their inwoven law-forces of differentiation, — demonstrating in the actual operations of nature,
that there is a power determined in an intellectivity,
as one force determines or modifies another, appearing
in the types of elemental molecules or atoms, in organisms, in functional forces of differentiation, in their orderly successions in time and place, and in all the correlations of forces in nature and life.   But all these, in
the broadest conception which can be given to them, cannot, without confusion of elemental facts lying within
the consciousness, give a wherefore, in the *moral feeling*, why the objectiv-faciency should create, why the
intellectivity should plan, design, functionalize so as to
bring forth in order, and provide for the continuance and
action through vast ages and eras of a creation unfolding to higher and nobler activities.   A still more recondite, yet openly apparent and essential element in man
— essence in God — is necessary to complete the entire
and perfect ideation of the forces at work to produce a
normal and intelligible act. I. i. 17, 36.   Actuous, objectifying power, and intellectivity, in and of themselves
give no love, or other sense or *causative love* of gratification for creation.   Why create ?   No inducing, attractive impulse in man, no causative love-movement in
God, no inducing motive-causation, as the causative
wherefore in a causative end, without which man is imperfect and God is without that coördinating essence
which is ascribed to him by the general consent and action of mankind, in all superstitions and religions, and in
all philosophies in which the Affections are recognized

as elements of the animate natures, and without which
there is no source of cause for the affections, desires,
hopes, joys, gratifications — loves, in animals, man, angels,
Redeemer, God. The Actuosity which goes over into
objective creation and into providences in the geologic
constructions and historical movements must be found
coördinated in causative love. It will be seen to be
deeply imbedded and inwoven in Commands, I. ii. 19, and
in the historical dispensations. I. vi. §§ 35, 36, 41, 33, 34.

11. Such a determinate Actuation, determined in a
causative Intellectivity, acting through and from and for
a causative Love, cannot be a spontaneity, for the Love
is at the end as well as in the beginning, and accompanies
the processes to this causative end. To be a determinate
Actuation, so to determine, there must be coördinate
powers — potential forces, I. i. 8, to determine to begin
— to pause — to change and break the series of causations
already at work — to do in conformity to the ordering
and adjusting of the Intellectivity for the introduction of
higher and other causative motive-love as the prolepsis
of the movement shall roll on towards the Final Cause —
the causative end. I. i. 15, 42. The Actualization is a
power, force, to do objectively, and so the Intellectivity
must be a power, force, to impress its designs, forms, intelligible functions on and by this Executive Force. To be
an active, causative Intelligence giving variety of form for
organization and endless variety of differentiate functions
in nature, in animal and vegetal life, in appropriate successions, there must be choice in means and intermediary
ends, and this is determinate Freedom, and this is incom-
patible with the inevitable necessity which must rule
development in matter and spontaneity in mind. I. i. 18.

12. Where, in man, the affectional character prevails,
with a predominate love of goodness and purity, and a

profound sympathy, which the world, in its animalistic
and human activities does not realize and reciprocate, and
the intellectivity cannot fathom the mystery of guilt and
goodness which exists in the individual human heart and
which makes the diversities and antagonisms of society, and
the Self cannot legitimate them in any consistent theory of
the universe, its rhapsodical intellectualizations will be
Mysticism — a love of holiness without the formal system
to legitimate its *feelings.* § 2. Lower down, in unauspi-
cious organizations, and in the fierce sympathies of epi-
demic *moral* hallucinations, as they are inaptly called, it
will be destructive and intolerant Fanaticism. It will be
apparent, upon an analysis of the Consciousness, especially
by the more advanced portions of the race, that this Love,
thus capable of attaching itself to the most dire fanaticisms
which have covered the incarnadine earth with slaughter
and desolation, and to goodness and purity which have
builded monuments of personal worth and integrity amid
these very desolations, — that this Love is an element of
its fundamental nature, and that its perversions are in the
organisms of the autonomy, and in the sympathetic influences of society, I. i. 34, and yet that a Self capable
of going out and over into actuation and of intellectualizing only, is a truncated and imperfect nature, and it is
imperfect in that it has no sense of gratification — no
motive-love in aught to be gratified. And the diversities
in individuals and races, and the antagonisms between
animalistic appetites and human activities, and between
the spiritual life and these animalistic and human impulsions, clearly manifest the imperfection of these races in
their individual members as only to be complemented
and made perfect in a higher coincidence or correlation
of the Love through the determinate Intellectivity carried
over into predominate actuation. It will therefore follow

that this Love — this affective element, inwrought and
inwoven in the composite organization of man, and placed
with correlating tendencies towards his animal and human
natures, and towards intuitive and ideative truth, I. i.
35–37, may be swayed towards either, or among the
various species of gratifications inwoven in the passions
and appetites of its animalistic and human natures, and,
there, may be compelled to elect between these various
gratifications, as well as between them and those duties
and offices leading to the gratification of the love of Truth,
Purity, and Goodness. And the philosophical contingencies of life, I. i. 13, are constantly, in higher and
higher alternatives, presenting the necessity of choice —
the easy descent or the moral conflict of the ascent. Thus,
throughout the whole of the complexure, in the web of
every life, in the mere motives of human prudence guiding
to this or that line of human personal conduct, and in the
conflict between these and the moral sense of responsibility, these alternatives are made to constitute the elective
contingencies of human conduct. Where in the peculiar
constitution of the individual, in his moral idiosyncrasy,
the love of truth, goodness, and purity prevails predom-
inately, and the mind is turned to the fundamental philosophy of life, it may lose or may not have gained the
thread of the system which will lead it to a clear standpoint for beholding the movements of the forming forces,
but it may retain, in the purity and freshness of this its
native characteristic, a lofty and holy Faith in humanity
and God; and here it is seen that this mind becomes
mystical, and its faith is the faith of Love, not of the
intellectivity, but through the whole of the processes
instituted by such a Self the *affection* in its higher appetencies is the attracting and controlling cause of its action.
And here it is seen how the men of greatest love have

the highest moral faith, yet liable in lower forms of organic growth to the orgasmic exacerbations of fanaticism. This correlate in man, essential to the fulness of his character, essential to any ideation of Morality in man or Deity, in the very necessity of our intusception filling up the content of a full and complete personality, compels the ideation of the coördinates in the All-mighty, All-wise, All-loving God — and that he can have no other attributes — all the other nominal attributes being resolvable into these, and these he must have. As the Self, thus moving through the phenomena of nature and life, gathers fact by fact, and sees that each phenomenon must have its special underlying force, and that these forces, when seen as physical forces, resolve into three fundamental bases, and when viewed as moral forces they are the accompaniments of the triplicate movements of the Self, both in its organic functionalizations and in its conscious autopsic action, it gathers them into a synthesis of triplicate correlate elements in the Self, and it cannot escape from the synthesis of these as coördinations in Being — God. When this analytic synthesis is declared, and the Self, by its own reflex intusception of itself, discriminates the animal from man, the animalistic from the human in man, and the human from the autopsic spiritual life, the discrete separateness of these elements and essences, as different yet as sympathetic and unifying forces in their triplicate conjunction, becomes a necessity of thought in the stringencies of both the Formal and the Moral Logic. I. ii. 19 ; iii. 28, 29 ; *ante*, § 5.

13. As there must be a determinated Actuation in a determinate Intellectivity, there must be a motive-cause to determine this intellectivity, and these are necessary to produce the various orders of creation, including man in his triplicate consciousness. The ideation of a morality

binding man to man, and man to God, is impossible to
thought without a coessentiality of Love — a love of
order, fitness, goodness, purity, both in man and in God;
and the same moral love of these inwoven in the conduct
of human life will also bind nature "fast in its fate" of
cause and effect in any system of moral responsibility.
I. ii. 19. Without order and fitness in the fate of physical cause and effect, the moral correlations of human life,
the provisions for raising the ever-renewing circumstances
on which hinge the responsibilities of the race in the use
and misuse and abuse of all of its animalistic and human
and autopsic powers and of the causations in nature and
life as the means and the evidence of their discharge or
neglect, would be wanting. The physical forces and *facta*
are but the instruments, the various means and furnishing
various modes by which the moral demonstrations of life
are made in vice, or guilt, or goodness. Man cannot
murder without the immanence of cause and effect in the
physical world, and these blend and inosculate with the
Moral Forces. The blow which is struck with a club is
called in common language physical force; when it is
struck with the fist, it is called muscular force; but the
club is only a prolongation of the arm by the addition of
a new lever : but the initiate forces which adjusted the
organism of the brain to the muscles of the arm, and
instituted the initiate causation by the influence of a
motive gratification (here the malignancy of a perverted
love), and informed it with mode, and selected means of
action, and went over into actuation, are moral forces.
The portion of the brain through which these forces were
adjusted for the terrible blow, which in itself was the
equivalent of many pounds' weight of force in any other
form of concreted or evolved force in any method of
measuring or weighing forces, — this portion of the brain

evolving this force, counterbalancing these other forces, will not weigh, probably, an ounce. So Charity and Fraud can only manifest themselves in the physical facts of nature ; and true charity, when it can, must deal in things substantial, and what it gives from love, the thief or cheat will take from a love ! The temple is the physical manifestation of piety ; the stately pile, of ostentatious pride ; yet these conditions of the Self, which so constantly work out into overt life, are not to be measured by the outward show of things, for there is pride in the alleys and virtue in the by-paths, and the widow's mite presented a wealth of soul which would have given a banker's noblest gift, but, throughout, the physical facts are the *counters* of the moral life. The moral and the so-called physical forces interweave at every step of the human and divine economies, I. iii. 25 *e* ; ii. 19. With man it must ever be fraud or force or sincerity and love ; and these can only be demonstrated and received and appreciated through moral, concreted into physical forces, which use the physical substances and powers, and which will always come in as the counters or symbols of that faith which is always shown by works ; — yet these can only be done,' and worthily and improvingly received and used and not abused, as the position and relations and correlations of the moral agents in the dependent and proleptic order which the Almighty has instituted for the moral dependencies and unfolding of the ages. " There is a real connection between natural and supernatural truth, and between all truths, because God, in whom all truth has its origin, is essentially one." — *Weninger*, c. iv. § 2. And what is true in the abstract must be true in the concrete, and nature and life unfold in actual truth. I. i. 8–10.

If the ideation of Being as the stable essence of phenomena is Matter, then the material necessity, υλικη αναγκη, is the philosophical product; the mind, in its rationalizing processes, being unable to escape from the invincible chain of actual material cause and effect, for it cannot posit Freedom as a product of its intellectivity otherwise than out of matter — out of the rigid order of causes and effects, nor can it be found in its primary causations as physical cause and effect. §§ 3–12.

If it is an ideation of Being as possessing only Actuation, or Intellectivity, or Actuosity and Intellectivity, then the human mind cannot escape from the labyrinth of Metaphysical Necessity. The unfailing Actuosity will work and crush and move on forever without intellective design or motive-sentiments for the attainment of an end in the future ; the chain of logical, intellective argument is the same irrevocable and unalterable course which carries on divine and human affairs ; and both are emotionless despotisms, relegating the affections, of every kind, from all source of origin, from all source of cause, and crushing them out in the inflexible tyranny of the logical Intellectivity and the actuating Causality.

But every act of absolute differentiating intelligence, logically, in the presence of motive-causations, necessitates a determinateness in each act in some wherefore — to what end — to some causative end implicated in the beginning. I. i. 15 ; ii. 19 ; iii. 28, 29. It is the end, in some gratification, love, which attracts, induces, infecundates, and enlifes the Intellectivity to form designs, devise means and ways, and select time and place, I. i. 1, 18–20 ; ii. 3, 11–19 ; iii. 1, 2, 9, 10, 20, and deliver them over to actuation. Without such conscious determinateness, limited intelligence, as the unnormalated intellectivity in man, as mere intelligential instinct in animals, will be impulse, spontaneity, — it may be, working to intelligent

ends, but the act, in itself, is not determined by its free
and independent intellectivity, but by its spontaneous
force. The unreflective Self has not yet cognized its
powers, and *taken possession* of them and guided them
by its more or less clear autopsic determinations; nor can
man normalate his life and arrive at clear self-consciousness, save in the supereminence of his autopsic consciousness, separating his animalistic instincts and his human
desires, and subjugating or controlling them with a conscious mastery. In man, determinate reasoning and
spontaneous psychical forces may be combined, in infancy
is to a great extent their natural combination, unfolding
in the gradual normalation of knowledge and reflection.
Thus there is, more or less, "a divinity that shapes his
ends, roughhew them how he will;" and yet within his
allowed circle he may exercise that determinate intellectivity, and by exercising it among the correlate powers
of the Self, in the orders of life, increase his allowed
circle. Every determinate act of intellectivity must have,
and implies a motive-end for its determination. That
motive is not found — cannot be found in its own strict
logical processes, except in those matters which are
wholly indifferent to any moral end. I. iii. 29; vi. 33,
34, 41–44.

Man, in the exuberance of his consciousness, must
affirm his own power to do, to act, or in doing, acting, — to
intellectualize designs and to actuate them in nature and
in life, — and his affections — his loves in various forms
of appetencies and towards various objects and pursuits.
One fits to the other. He finds physical nature adapted
to his actuation in diversified forms; it responds to the
designs, forms, figures, ways, and means of the intellectivity; and in all the animalistic appetites and human
wants and desires it furnishes the elements of gratifica-

tion and for choice — election. I. ii. 19. And the correspondences are seen as those of correlations. And it is not for man to say whether he will affirm his power of actuation, of intellectualizing and loving, or not. He cannot but do it. It is in the direction of this love, in man's elections between lower and higher, and the higher and the highest object of gratification, that his responsibilities are hinged and his progress is allowed. And this gives the moral end, — and thus it is seen that the End as Causation is inwoven in the processes and in the progress to the End. I. i. 15. Without the stabilitation — the regularity of cause and effect, and the flowing and becoming of these effects, and their adaptabilities adjusted to the moral forces in the Self, and without the separations of cause and effect in time and space, giving room to the philosophical contingencies in nature for the determinate actuation of man, there can be no conscious exercise of responsibilities — of the moral forces in the Self. They are adjusted each to the other, and there is no theory or fact in nature and life, or of the correlations which spirit bears to the physical world, by which these moral forces in the Self can become physical forces in the actuation of the Self, — as a blow by the hand, — how the Self can move static force, how it can control or set at liberty centrifugal or centripetal forces in itself and in the physical world, unless upon some identity or kinhood of intimate correlations between all Forces. The Intellectivity is now, so far, seen as a causative force : it is seen as a conscious causative force. With a love in some end of conscious gratification it is a Moral force — *they* are moral forces, in their correlation as they must be in their coördination. I. i. 15, 36 ; ii. 11–19 ; iii. 1–3, 5, 6, 15, 16, 20–31. In the Absolute Intellectivity the moral end of actuation cannot be found in the pure insistent truth, for it is seen as an eternal abstraction. and as it were only an absolute form — two and two are four as an eternal abstraction, etc. I. i. 24. There is no motive-end in it. I. i. 35. It is the mathematical logic without any content of life in it until it is supplied in actual existences by quantities and qualities. I. iii. 29, 28. Nor can this moral end be found in the naked and pure acts of the Intellectivity. if such are conceivable as normal acts, which produces the determinately selected organisms for the various autonomic organizations of nature ; for although there is no known instinct of gratification in the blindly intelligential workings of the forces which build up the vegetal life. yet when *functions* of animate life appear in the instincts and their correlations in nature are supplied, I. i. 30, a sense of gratification, a want, desire, love of somewhat, and in most animals many such are introduced. But, at every step, the question must be asked. wherefore ? to what end ? If to a purely logical, mathematical. transcendental, intuitional end in the intellectivity. as such. the Self is only in the logical, mathematical, transcendental, or intuitional state, and cannot escape from the rationalizing processes of the mathematical or metaphysical necessity. There must, in the very virtue and moral necessity of things as correlated in the loves. hopes, fears of man, with — not distinguished against and as independent of the intellectual order, but with it, be an end in some motive-appetency — some attracting causation. to which determinate intellectivity is a means — a designer and framer of means and modes for the use or attainment of the end ; but it is the End which is sought. and the end furnishes the motive-causation throughout the intermediary means and ends to it — the Final Causative End. If, then, to an end which may, as in man and speaking as man, gratify this,

that, or other appetency, then there is a choice, a freedom
in the election of the end and in the selection of means,
out of many, which may be appropriate to the end. In
man these are the elements for giving play and scope, and
growth and comprehension, to the intellectivity in its ever-
recurring and widening complexures in life for the exercise of its autopsic selections of the end to be pursued and
for the dominancy — the fact of the complex will going
over into actuation. In God, whatever this essence of
motive, this ormaic *movement* to create is, it must be a
simple single coördinate force, and as such coessential with
the Intellectivity which furnished the Divine Ideas. As
a movement-force, it is a coördinate power, — and it is
Love or hate! These coördinated with Objectifying
Power and carried over into objectivity, is Creation.
I. i. 8–10. Identity cannot produce diversity, I. i. 8,
and a motive-end in the personal Self is a necessity
of thought to any ideation of moral actuation. Actuation in man and God is a positive force, and as such
is modified by intellectivity, as the intellectivity is in
turn modified by the Love. Then this intellectivity in
modifying another force is only conceivable as an intellective force, and the objectifying force and the intellective
force modified by the *attracting* motive-force, the Love,
thus necessarily seen in a creation where there are moral
means and moral ends inwoven in the movements of its
life, it too is seen as a force, and in their coeternal coördination they ever are and remain Moral Forces — in
their uses and ends. While this last causative force, the
Love, in Deity is, in itself, *speculatively* separable from its
coördinating forces, and so is seen to be a simple essence,
yet in God they are always in coördination, in man it
may be as variously functionalized and complex as the
loves of gratifications inwoven in his animalistic and

human natures, I. iii. 15, and in the correlations to and
of outward natures and in other selves adapted to these
gratifications — in the sympathies and antagonisms of
these very repelling, attracting, and redactive forces.

14. Material and Metaphysical necessity both arise
out of subjective conditions of the mind — the Self, and
are both alike. They are both Rationalisms — processes
of the mere truncated Self in the uncorrelated and unnormalated and *unconditioned* action of the Intellectivity.
They differ only in this : in the former the mind refers
to matter, more explicitly, the logical necessity which it
finds in itself — in its formal chain of cause and effect ;
it borrows cause and effect from nature, and in its partial
and truncated philosophizing refers the proper moral
necessity for cause and effect, §§ 4, 5, 13 ; ii. 19 ; iii. 25 *e*,
back again to nature as its original, self-inexistent, and
eternal law of necessity. Pursuing its rationalizing
analyses, it perceives the facts of causes and effects in
their successions, and it can only see *a posteriori*, in its
mere logic, a linked series of cause and effect — and this
becomes, to it, the Material Necessity. It perceives that
matter has no inexistent and independent motive-force
to break up the catenation, or rather the unarticulated
ongoing of cause, and hence it is effect from cause end-
lessly. In the latter, the mind, consciously or unconsciously, occupying a purely rationalistic position outside
of and beneath or on this side of nature, in its own chain
of exclusively logical reasoning, can see no beginning
nor ending of its ideological processes, which in one direction must be cause, and cause on forever, and in the
other direction be effect, effect interminably. For in
ideological processes of this kind the God at the beginning of this chain of effect from Cause is only an assumption, — there is no intusception of conscious cause-

working forces, and there is no legitimation for the origins of the moral forces, the functionalized organisms, and the correlations of the moral forces between themselves and between them and the physical forces. This process has gone back until, oppressed and exhausted or bewildered and mystical, it says here God began as Prime Cause. Nature and God are thus bound up in the chains of our logic, and there is no motive-love in means and ends to set the Self free. This is but the exclusively rationalistic analysis, excluding the cognate elements of the Self, and therefore not finding them in Deity. It is an imperfect rationalistic synthesis of nature, omitting the coördinate elements of a true system. I. i. 38–40.

15. To break up this chain of Necessity, thus reduced to its simple fundamental derivation in the logic of the Intellectivity, there must be an appetency, a choice, that out of a number of means to accomplish an end — an end in the love of some action or end, the intellectualized Love shall select, and the Self at once ascends above nature and life and finds the causative forces moving into and actualizing nature and life, and thus gets objectiv-facient power, Intellectivity, and Love as coördinate forces. The Actuation, the Intellectivity, and an attracting Appetency correlate, in man, constantly with each other in the normal movements of his autopsic life ; in Deity the ormaic motive-force must coördinate with both the Creative-force and the Intellectivity, or the mind cannot escape from the material or the metaphysical fatalism. If the Actuation is supreme and acts *sua sponte* — from its own inner developing force, it is a blind power working without intelligible means to *un*intelligible ends, at least to ends which have not intelligential final causes ingrafted proleptically in their constitutional correlations, and which are to fall out in a succession of intelligible harmonies

and advancing perfectibilities for maintaining and attaining the orders of the successions and for the accomplishment of intermediary and ultimate ends and an ultimate end. There is here no prolepsis inwoven in nature or in mind, and even the material necessity or *systematic* catenation of cause and effect is destroyed. There is no purpose-end, no final cause, nor any sensible or significant correlations of adjustabilities or of system in any fact or symbol of nature, or any combination of organisms, functions, or psychical forces inwoven in them to accomplish a purpose or to fall into an intellective system. If Intellectivity is alone hypothesized, as a mere abstraction, as it is in all philosophies and theologies, then there is no actuous objectiv-faciency, no working creative force, but simply an eternal contemplation and dreaming. This is the *nieban* — the eternal repose of the ancient and still prevailing theosophy of the East Indian Buddhist. The moral element, the motive-essence, the causative love, without which there can be no Morality in God or Man, I. i. 35–37, etc., is wanting to co-act with the Intellectivity, and give to the necessity of thought the ideation and the impulsion of gratification — a purpose — a final motive, attractive cause inducing to the acts of creation and to be gratified ; but all the Forces, thus united and conjointly acting as Creative Forces, are thus attained analytically, synthetically, and in a Personality, redactively, when the whole content of the Consciousness, I. ii. 2, 14; iii. 9, in its triplicate correlations, are taken up into the transcendental position. No philosophy of Mind, in its broadest acceptation, no intelligible system of the cosmos is fundamental until there is a final analysis and discrimination in their ultimate bases of making — doing — of intellectualizing, and of loving in their correlations in existences, and in their coördinations in Being, and these are synthesized and their unitary and unifying bond of union is found in a Redactive Personality. In existences, this synthetic union is found in the adjusting correlations which bind the Feelings in animate nature in their actual correlations of life and in the moral associations and antagonisms of human society; while in Being it will be found in the clear beholding of the eternal and harmonious coördinations which give system to all the systems of successions and their gradations.

16. A triplicate of Essences will be found — are thus found — and only can be found in the complete analysis of the Consciousness, that undoubtable matter of fact which is the lawful province of human examination, wherein is seen that image whose content gives the image of Him who created, — let us make man after our own image, — and this can only be by the Intelligibility of the Creator inwoven into man — the symbol. This is the Scriptural authority for instituting this process, while to the philosophical mind the conviction must flow from its own independent processes, and both processes, fundamentally understood and wisely applied, will unite and harmonize in the method of uplifting the human solidarity from the obscurations and environment of its natural condition, and give discriminate aids to the movement of tribal and historical progress. In the union of these processes, if both are true, they will irradiate each other, and the accumulated observation of the facts, collected from the movements of the geologic eras and the historical unfoldings of the moral bearings of the races to one another and among the individuals of society, will furnish the evidence of new inweavements and concretion of higher forms of these forces into the life of humanity. The test of a rigid psychology, verified by a rational

11

transcendentalism, based on a recognition of those fundamental forces, must be applied to the processes, which
can only be done in all the facts of the consciousness.
This cannot be done in a simple Rationalism, but requires
the Intusception, by which the Actuation and the Affections shall be recognized as fundamental as the Intellectivity — yet only through the Intellections. Christianity
teaches the doctrine of a triune God, wonderfully preserved in the transmutations of language (see the change
of *hypostasis* in Greek into *persons* in Latin and English)
and the changing vicissitudes of philosophic schools and
theologic dogmas, and that man in the renewal of his
knowledge may attain by a new movement of life — the
New-Birth into the Spiritual Life — a nearer likeness to
• the Divine Life. As the Ascent is made, it will be seen
that this New Life is an inweavement, a proleptic and at
length a conscious exercise of the Love in the pursuit
and attainment of Holiness. § 24; vi. 18, 40. The
entire content of the Consciousness, in its natural elements, must be analyzed and synthesized, and, ascending
or departing with this clue, by this Method, in the pure
light of the Mind, the explication will be clear and definite,
and the synthetic correlations of these triune forces in
man and of the coördinations, from everlasting, in God
will stand forth in their sublime simplicity, and their
capacities for the universal diversifications. In adopting
this Method the universal rule in Science is pursued, and
all the facts are sought and admitted into the calculus.

17. Whatever is a compound may be analyzed. And
whatever is simple in its essence and yet capable with other
simple essences of forming diversifications, differentiations,
may be observed and noted in their phenomenalizations
into these diverse differentiations. And these analyses
may be intellectually — intusceptively reconstructed. The

very process of carefully and fully conducted analysis is reconstruction by the synthesis, simultaneously employed, the only intellectual and moral reconstruction of which the Self is capable. I. i. 38–40. In nature this is the reconstruction of elements into the subject, or the object thus found by the analysis, in such subject or object analyzed; and in the domain of spirit it is the synthesis of elements found and essences attained by its own self-analysis. Spirit, in its essences, the Self may only know in its phenomenalizations, and matter it can but know in its underlying forces of concretion and movement. I. i. 2–4; ii. 2, 11, 13–18; iii. 1, 2, 9, 20. The philosophic spirit, for thirty centuries, has exhausted itself in pursuit of these underlying identities, and, although always implicitly affirmed, the gulf which separates them from our discriminate cognition remains unpassed, except in the more accurate definitions and classifications of their phenomenal facts, and these as constantly converging to a knowledge of their primitive bases of origination by the establishment of their mutual correlations and essential differences. The cognizing agent, the Self, deals with their phenomena as actual cognitions, and, in virtue of its syntactic synthesis, notionalizes the underlying identities from and by the uniformity of their respective phenomena, I. iii. 20; for, cognizing phenomena clearly and distinctly in their essential differences they become capable of classification, and are thus referrible, in their uniformities, to the distinctive *somewhats* — noumena of their sources. This cognizing, doing, loving Somewhat, giving forth these special phenomena, is a complexus having triplicate activities in its elemental nature; for this Agent-Self, in its consciousness, may know and desire or act — it may desire and not act, but it cannot consciously and determinately act without knowing and desiring, however pain-

ful and trying it may be under certain given circumstances to act upon one desire or preference rather than
another or others : it may think, intellectualize about
what it may know, without acting or doing overtly in
reference to it ; in mania the actuation is frequently dis-
connected — discorrelated from the intellectivity, and acts
without and beyond its normal influence and control, as
is seen in some wards of the Asylums of the Insane, where
fierceness and frenzy prevail ; while in others the monomaniac exhibits the predominance of some fixed current
of thought, and in others the exacerbation of some affection. This furnishes distinguishing evidence that in each
of these distinctive departments of the Self there are
laws and forces of separate intensification, which, when
continued, end in breaking up the normal correlations
between the three, and mania and monomania and fanaticisms are the legitimate results. While this is so, this
evidence becomes more confirmatory of the facts of their
separate and distinctive differences, and that they are
controllable and controlled, in a proper normalation of life,
by the autopsic powers of the presiding Self, and that
there may be a growth of the separate powers in the
psychical organization, and some may be suppressed and
others intensified and made more active. I. iii. 3–6, 9,
15. These facts further show that the Self, in the complexure of its autonomic and psychical organisms, has its
laws of positive growth, or of capacity for so altering,
changing, and modifying these organisms as to be susceptible and capable of a more complete and open manifestation of its own identical, spiritual, solidaric powers. In
the historical and philosophic movements of life this is
the mentalization of mankind ; in the still higher exaltation of life it is the evangelization — the life-integration
and inweavement of the holy and depurated Love. The

Self, when directing its investigations to man, in and through its own self, educes the science of psychology, and necessarily includes its empirical and rational processes; in its Intuitions it gains the Insistent truth, I. i. 35; and by its ideations it ascends to the Divine Ideas and their systematic correlations for uniting the whole into an orderly movement for the worlds and the existences therein, I. i. 37; and it finds all these bound together in the laws of a Proleptic Morality adapted to the government and training of autopsic agents living and moving through the grand panorama of the Divine Prolepsis, I. i. 36, 42; ii. 19; and when directed to nature as well as life in all its vegetal, animal, and psychic forms and autopsic manifestations, it is a psychologizing of the whole through to their Initiate Forces. It is the interpenetration, the intusception, in and by its own conscious triplicate activities into the material and psychical phenomena of nature and of life, and the cognizing of its own spiritual powers and the ideation of the Fore-plan and the gradual and proleptic uplifting of the Self—of all selves—in their times and places and lives of this unfolding ordination.

18. The conscious perspicience of the subjective content of the Self in the action of its orgasmic forces and of its own autopsic domination is Psychology. The phenomena of the Self, possessing uniformity, each in and of its kind, and thus each its own inwoven Intelligibility, speak to the Intelligence on this side of nature and life, and are thus capable of classification, of being bound together in the synthetic fascicles of their root-forces, and each class can be referred, thus-wise, to their respective somewhats — their *substans* — their certain and definite *noumena*. I. i. 3–5, 24, 39, 40. A classification is necessitated, and as necessarily so as of the *predicates* of matter, and this must end in the notionalization of an

underlying *substans* of force, for each radical difference of phenomenalizing facts. In the analyses necessary to this classification, the facts have been slowly accumulating in the conflicts of systems of philosophies and creeds of contending theologies. •The interpenetration, the conscious, intusceptive examination of the content of the Self, and of the *effects* produced upon its correlate organisms by the Self, and by the objective forces in nature and life, will be found to follow three distinct and well-defined modes of procedure. I. iii. 20.

a. The phenomena of the outer, exterior sense-world — outer to the central Self — as they modify and affect the Self in sensations, *registrate* their modifications in and through the appropriate organisms. I. iii. 3, 5, 11. When the Registration is made, it becomes subjective ; it is a modification of the conscious Self, and when it is necessary to recall it, or when it is recalled by direct effort of the Self or by associative causes, it is only by being *re*-enlifed by the Self consciously or associatively, recalling, reproducing from on the other side in and from the Self, an *effect* similar to the original sensation. It is in a certain sense telegraphed back from the Self. iv. 25. This is done directly and by associative causes — causations. II. vii. The Self, in this *re*-cognition of the *sensate*, does not perceive the original object which produced the sensation, but this subjective-object — the sensate reproduced in its own organism. This is called Imaginate. I. v. 26.

b. The activities which are at play within, in its own interior psychical operations in and through its various organisms, are to be observed. Each act or operation of the Self, so far as it can be seized and held by the self-conscious consciousness, becomes an object to it, that is, these acts become subjectively objective. These activities or operations are seen to fall into three well-marked and designate activities of phenomenalization. I. i. 17–23 ; ii. 3, 14, 16 ; iii. 9, 15. Still observe that two classes of these manifestations are spontaneities, and that the third, the intellective, may by intensification become automatic and act in the nature of the spontaneities — as in ideational monomania, as also in revery and gossip.

c. The self-conscious Self, standing at its own centre of organic complexures, catches these senses-registrations as in a, and the psychical activities as in b ; and from that point it acts self-determinately, outwardly, in its actualizations from its intellective selection of means, · modes, and end, in and for some sense of a gratification : it there, in open vision, beholds the Insistent Truth, and catches the Divine Ideas and the Proleptic Morality.

19. The intuitive intusception — the psychological action of the Self, in the complement of its triplicate nature, is introverted from the material world inwardly to these realms of Intuitive and Ideative Truth, yet with the symbols of the material world always present in their registrations and with the facts of its psychical life to be relighted with the light from within. Man, the psychologist of nature and himself and of other selves, — Man, the intelligent Seer, standing, as it were, in the nexus of an illimitably diverging hour-glass, the two cones of which may stand for the two worlds, one of which presents its facts and objects in Sensations from without, and the other from within, — correlates and joins the two, and the true progress of the Self is obtained by this union of the Sensible and the Supersensible. It is the marriage of the Actual and the Divine. I. i. 19. Man thus conjoins the sensible and the supersensible, and reconstructs for himself the divine prolepsis and works, more or less consciously, in his time and place in the onward, ever-rolling movement.

20. Common language discriminates between inorganic matter, organisms, and functions — the orgasmic forces which vitalize in and through these organic functions; and philosophy recognizes the differences between intellect, intellectivity, and intelligence. Intellect in man is the functional apparatus through which the Self learns and cognizes; intellectivity is the learning, cognizing agent; intelligence in man is the particular or general result of any or all the acts of the intellectivity, and, in man, neither the one or the others are conceivable as human and active agencies without the whole. In God, intelligence is the pure absolute omniscience; it is above all organization, for it produces, and, in its coördinations, functionalizes all organisms. It is always, as omniscience, without a past and future, yet always has a past and future, a time and place in the manifestations of its actualizings in creation and in the observations of intelligent existences in their proleptic orders in time and place. But the ideation of an omniscient intellectivity in God, thus assigning his Divine Ideas to their proleptic orders in time and place, necessitates a coördinate Creative Force to give a material and life-content to these Divine Ideas and to objectify them forth into their stable immanence of existence and action in *his* times and places. The Insistent Truth, the Geometric Laws, the Divine Ideas and the Proleptic Morality are but empty abstractions, in man or God, without their material contents and working functionalizations. Thus matter, organisms, and functions are redactively symbolized, vitally actualized, and the ends and aims of their existences intellectually teleologized. But this furnishes no motive-element — vitalizing power of moral life; the objective force, that force which is correlated with or which is the spontaneity of anger in man,

the mere doing actuous force is a blind power, in and of itself; and while the Intellectivity, in determinate acts of symbolification and of teleology, must give form, modes, and means to the actuous force, yet to what end and wherefore shall these move, if not to a gratification — a Love, an ormaic impulse, and therefore an essential, attractive, inducing power. The moral teleology is wholly wanting. I. i. 17–23; ii. 2, 3, 11–15; iii. 1, 2, 9, 10, 20, 25, 28–31. The Actualizing force is articulately guided by the Intellective Designer which is ormaized — coördinated in the Divine Love, and thus are given the threefold Essences of a Divine Personality.

21. With these views it is, for the present, affirmed, for the purposes of Positive Science, that,

a. The ontologic side of the objectiv-faciency in Deity or in man is the Creative Power of putting their acts, action, actualizations over from the Self. Man is, in his sphere, a creator co-working in the movement of the Divine Prolepsis. He makes the daily and artistic forms which gives labor, taste, and skill their employments, and he gives them a content and powers from the elements furnished him in the materials and forces of nature and of life. § 11. The phenomenal side of this Force is, conversely, objectivity in action — projective power — tangential projectivity, — and this is that coördinate power by which nature is not a Pantheism, but is set over objectively from God.

b. The ontologic side of organisms, in forms, functions, correlations, is *in* the Intellectivity, and these are shaped and differentiated in ways, means, and forms of action by the Redactive — the Intellective Force. I. i. 40. The phenomenal side of the Intellectivity is these forms, organisms, differentiations, functions in their formal redactions as impressed from the intellectivity in the Divine Ideas, and actualized into life and nature.

c. The ontologic side of wants, desires, affections, &c., is Love ; and *e converso.*

d. That *a* in God is Omnipotency — Objectifying-efficiency.

e. That *b* in Him is Omniscient Intellectivity, arranging the appearance of his Divine Ideas and their intellective correlations in time and place.

f. That *c* in Him is the creative — causative Love — the Attracting force for all spiritual life. It is the Beginning and the End — the prime Efficient and the Final Cause. I. i. 15; ii. 19 ; iii. 29, 31.

22. Such are the elements of all knowledge, the method of attaining which and comprehending all objects and movements in nature and life is Intusception. This is the process by which the Self enters into the forms, organisms, functions, forces, wants, desires, passions, appetites, and all activities and movements, and flowing, growing, moving in, around, and through them with the consubstantial forces which created them, which produced, is producing, altering, infusing or has infused life, energy, motion in them, and so and in such modes has evolved, and, in their objective immanence of forces, is evolving their constant phenomena — and in this personative and representative method this Self attains its cognitions and its philosophy. No idea, law, principle, activity, or affection is understood, is comprehended in the significant sense of this term, until it is in some sort infecundated, enlifed, actualized by the Self in some or all of its trine activities, in greater or less combinations of each. In the ultimate analysis every determinate act will be found to unite in it the action of the triplicate elements. The intusception must be complete and exact, or the idea, law, principle, actuous force, affection, will not be completely and exactly comprehended ; and the perspicuity

of the comprehension will be as the completeness and
the clearness of the intusception. In making this intusception from the transcendental standpoint, it will be
seen that in the proleptic ordination of an Omniscient
Intellectivity, to which there is no absolute time or
place, § 20, but all is a Past-Future-Now, the *conscious*
Love was postponed in the proleptic order, and as this
unfolds, it becomes more open, clear, and intellectualized,
" renewed in knowledge," for actuation. Man learns,
and hence it is that every Self, in its ideations, forces,
and affections, as working or having worked in nature
and life, is, in some sort, a creator, ποιητης, a projector
of its own spirit into the subject-object of investigation,
and in this postponement of the Love, it, too, always has
it for its end of movement, as a causative end. Hence
the idea-discoverer so surely desires, as the prolongation
of his loving activity, to be an actor of it — to actualize
it — to symbolize it in word or deed, in poem, or sculp-
ture or architecture or evangelization of life — of all
lives. Hence, as minds expand in the mentalization of
historical movements in society, and the horizon of thought
and action unfolds the glories and wonders of the Mighty
Life which sways the universe in the harmonious systems
of his Intellectivity and under the sanctions of his unfolding Love, these minds are bathed in the light and
harmonies of this Love, and man desires freedom of
thought and action, that, in the limits of his allowed
circle, he

"May make the chalice of the big round year
Run o'er with gladness."

And thus it is that man, having filled the forms of nature
and life with his Spirit, would, in the noblest aspiration
of his philosophizing, attain to a position outside of
nature, fast by the throne of God, and then pass down

through nature, and, psychologizing its atoms, planets, suns, and systems, and all organisms, functions, forces, intuscept, comprehend, and with his Spirit infecundate and enlife the whole. He would aspire to God, and on the wings of his Love would lift the Solidarity of Humanity from the deep, dark depths of its animalistic and human depravities. The Intelligibilities from above are precisely correspondent so far to the aspiration of the Intelligences beneath ; and as the Intusceptive Self moves forward, it gathers in the volume of these Powers.

23. Sensations produce certain and different effects on the Self. The conscious Mind takes note of these, and the difference of Sensations is the means of catching cognitions of different things in their quantities, qualities, and their intensifications. While this is so, it is also in virtue of the differences in the organisms adapted to and correlated with the things which produce these different sensations : the eye sees, the ear hears, &c., not only in consequence of the difference of quantities, qualities, and intensification of forces in the things which thus *effect* them, but in further virtue of the powers which functionalize the organisms, and the perfectness of the organisms themselves. Thus there are organisms which are intellectual, others are affective, embodying the loves of our gratifications, and others are for motion — actuation ; yet all are capable of united action, and in all normal action — actuation — are united, and in a well normalated life are placed under the direction and control of the Autopsic Self. I. i. 33 ; iii. 9, 29. If all sensations were alike, there would be no appreciable difference in things, nor in the effects on the Self, and distinction and discrimination would be lost in one common platitude of sensation. If there were no differences of sensations, there would be no discriminations in effects, and no possibility, in us, of

reaching to the secondary or primary ontologies which produce them. The diverse Intelligibilities from on the other side must have diverse Intelligences on this side for their apprehension and their appreciation on this side. It is, to the Self, the difference of the mode or degree in which different things *effect* it and affect it, by modification of the external or internal senses, which is the foundation of its knowledge of sensible things, and of internal sensations. This is objective knowledge made subjective in virtue of the very effects and affects of the various organisms involved and concerned in the operations of nature and life. ·It is the knowledge of the modification of the afferent nerves and their correlate organisms in the psychical organization, or in direct communication with them and putting them into activity. The psychical functions only become cognized as such later in life, when a self-analysis has begun to be instituted, and the discriminations are begun to be taken between the knowledge brought by the external senses, the impulsions from the animalistic nature in the "war in the members," and the human appetencies and activities among and toward the things and persons of this life as correlated to the Self in this earthly theatre of action. As the discriminations are made, each class of these phenomenal effects are seen as entirely distinct and enclosed in separate sets of organisms; and thus it is that, as the progress is made, the war between the higher and lower orgasms is increased, until the one or the other conquers and supervenes, and until the revulsion from these comes and the Autopsic Self, in the plenitude of its deobscurated powers, in its pure love of Holiness, obtains and holds the mastery over all. It is manifest that he who is in the lower condition cannot at all, or only faintly and by degrees, as he conquers for himself, yet with extrinsic aids, I. i. 19, and makes the ascent, institute and maintain in his practical life the discriminations. The distinction between the Self and Objective nature is the early instruction of infancy, but the discriminations cannot be made in the latter cases until the Self loses much of its spontaneous impulsivenesses welling out from its orgasmic forces, and the Self from its own self — from its own centre of self-command can, more or less clearly, reproduce, and thus reënlife itself in the sensates and these soul-activities, and cognize them in the effects upon and from their particular organisms, and bring; from their spiritual sources, the powers to modify or govern them. By these means, that which is subjective in the organisms of the Self is made objective to the cognizing Self, as by like means, in the objective control and mastery over the orgasms, the Self becomes self-conscious of its mastery and domination, and of its super-tendencies to the Suprasensible — the Godward side of its existence. In all these processes there is a more or less active state of the Self; and a passive-active or active movement of the object or the subject-object, as the case may be; that is, there is in the objectivity, or the medium by which it is perceived, some activity — some movement-power by which some one or more of the sense-bearers are affected and perception in the Self is produced: reflected light, touch received or made, flavor, odor, venereal orgasm, gastric juice, motions of nature, &c., require a receptivity in the personal inspection in the consciousness to determine what they are; and below this in the animal kingdom, where the passions inwoven in the internal orgasms produce their respondent acts without the intervention of a consciously controlling Self, they are instincts, as in man this uncontrolled or uncontrollable action is automacy or monomania, and so are not autopsic cognitions and moral control. I. iii. 11. Yet down in these Instincts of the animal orders, it will be seen, II. iv., that, as instincts, they have the power to act, to do : the blind inwoven intelligences which direct them to the objects of gratification and to perform the instinctive acts wisely and well, and for the gratification which is the end of the instinct, and all correlately inwoven in the instinct — while things in nature are correlated to these instincts in animals, and to the conscious gratifications in man. § 13.

24. The Self makes a weary pilgrimage through the instructions and discipline of infancy and the burning passions of pubescence, before it arrives at the self-analysis which makes these discriminations and unfolds in the Self the self-conscious power of self-government and regulation of life, for the control and subduction of the passions and appetites and exercise of its faculties for æsthetic culture in matters of art, and taste, and for the discipline and education of life in subjecting these animalistic and human orgasms into obedience to the Proleptic Morality, I. i. 36, as in the ancient Stoics, or, in a more comprehensive system of thought and actuation, as found in some hypothesis of natural theology, as the commencement of a prolonged career beyond the bounds of this life, or as in a deeper and more terrible self-analysis, from its ideative standpoint fast by the throne of the Infinite Supreme, in which the Self beholds this life in its low and unworthy state, it finds its moral end and means of action, and moves onward in its own depuration to the attainment of the Ultimate Morality, wherein the Proleptic Morality is seen but as a scaffolding for ascending to the Love of God. II. viii. The Intelligible Love from on the other side unites with the Intelligent Love from on this side. The Marriage of Heaven and Earth. I. vi. 18, 49.

25. If an ontology is affirmed for the Insistent Truth or the Divine Ideas, this carries us back to the "eternal patterns which subsist according to sameness," as sup- .posed by Plato, and upon this supposition would be intuitable by the Self in their own objective light, in a manner similar to that in which sensible objects are perceived through the sense-organs. But this would give an eternal subsistence coexisting with God and yet not God. Insistent truth, as in any mathematical proposition, is but the abstract forms of quantities. I. iii. 29. The Ultimate Morality is the law or rather unity of the coördination of his coessential powers by which he is seen as All-mighty, All-wise, and All-loving, and the Proleptic Morality is the statutory, appointed law of Progress for self-conscious and responsible agents, by observance of which they may gain a higher spiritual, solidaric life. ⸜ Yet all these must in some way be con- sistent and harmonious. The human intellectivity has its laws of order — law-forces — of cognizing and ideat- ing, or it is disorderly and confused. So these intuitions are the reflections from symbols and the intellective force inwoven in the Self of the orderly processes of the Divine Intellectivity, after which it was "imaged"; — so the pro- found love, welling up in man, in its spiritual enthrone- ment, will have its correlations of order, justice, righteous- ness, and as these are broken or do not exist, vice, wrong, injustice, and sin, in the self and in society, is the conse- quence; and as the perfect correlations are restored, order, justice, righteousness, reappears in man and in society. I. ii. 19; iii. 28, 29. This is but representative of the coördinations which rule, as it were, the coaction of the Divine Coessentialities, and make order, justice; righteousness, but synonyms in the Divine Self. Thus in the organization of a man of a high intellectivity the

mathematical order or intellective intuition may appear in a dry, hard manifestation of these intellective processes; in another, the ideative power may appear, and be so imbued with the love of moral order, justice, righteousness, as to appear the very order and of the very elemental powers of the organization of his mind ; and so all these appear in the coördinations of the Creative Power, Omniscience, and Love as but the order of their eternal harmonies. And when man ascends nearest to these, he is likest to God. This is the Final Cause inwoven in the Prolepsis, I. i. 15, 42, and to become the self-conscious possession of those who work wisely, meekly, and morally obedient in the proleptic movement. If the Immutable Morality is an *abstract insistence!* then it has no ontology for effects, no causative powers, and is without any necessary connection with Deity ; but if it is essential as of the coördinate movements in the harmonies of the Deific Forces, and is thus and only *thus* ontologically causative, then, as these forces are gained and imparted in their purity and simplicity to the intelligent and loving coworkers on this side, man unfolds and deobscurates his solidaric life. Deity is seen as Essential Truth in Essential Life, and man only gains this Truth as he gains this Life. "I am the Way, the Truth, and the Life," was the embodiment of Truth and Love and Actuation into Life.

26. The clearness of the vision of the sense-world depends upon the perfection of the organizations of the outer senses and upon the contingencies of the circumstances under which the observation takes place ; the perceptions of the impulsions from the animalistic and the human orgasms, so near akin to those of the higher animal orders, in like manner depend on the orgasmic powers, as causes, inwoven in the organisms and in the-

perfectness of the nerves — the telegraph lines of communication in to the Self. Injury, disease, characterizing temperament, medicinal agents, may *effect* all these. These are wholly distinct from Intuition and Ideation, and the two — the outer world and the orgasmic sensations on the one side, and the intuitions and the ideations on the other — are seen from directly opposite standpoints in the Self; — the first, wholly from without and coming from the sense-world ; the second, as clearly coming from the animalistic and the human natures within the organic structure; the third is a simple operation of the intellectivity in virtue of its native, elemental quality or nature ; and the fourth is the cognition of Being — of a coördination of positive and coessential forces, Intellective, Loving, and Actuous, as intuscepted by the conscious autopsic movements of the Self in its triplicate consciousness. Face answers to face. Thus are given the correlations between the Self, in its triplicate elementary Intelligibilities inwoven by a divine creation into its constitution, and the Intelligence from beyond, in its trinal coördinations, weaving the web of the vast complexure.

27. As an ascending ideation, the following illustrations of the processes between the Self and its objects of intusceptive observation may make more palpable : — the immanent rock is perceived, which, in its qualities and quantity, gives to the attending, cognizing Self information of the rock — so far ; light is reflected, and it is seen and is gray, and the educated sight says it is large ; it is touched and it is cold; it is heavy and resists *in situ* — giving four elements of a somewhat in time and place. But what are the underlying causations which gave it solid extension ; what, which gives heat and the absence of which gives cold, and shows it only as an *accident*, yet always as inherent in some form ; what is

the attracting, gravitating force which holds *in situ*; what the light by which it is seen, and what is it which gives color; and what is the thing itself possessing these qualities, when they all are found in like manner in other things? Again, the rose may be seen and touched, but it gives out odor; in the one case the rock has an active-passivity, — it is held by the earth, and it holds on to the earth by gravity — attraction; its passivity gives a rebound — a reflecting surface to the immediate action of light, by which it is seen.; — in the latter case the rose has the passivity of the rock, differing in this respect only in degree; it throws back the light in various colors in virtue of its own organization breaking up the trifold elements of light into these various colors, and it is more immediately active in its inherent force of expulsion of its odorous particles. I. i. 3–5. The Self now perceives a horse, but it sees more than the inanimate statue; it sees all in the horse that is in the rock and in the rose, only differently organized : the horse extrudes from himself to· the perceiving agent qualities and activities still more highly organized than those of the rock and the rose, and this cognizing agent seems to grasp more readily these newer and higher and more open forces of the horse, and he really does so because they are nearer in open action to his own more conscious exercise of the same forces. Substitute a man of intelligence and love and actuation, in the full play of his perfect nature, for the horse, and the Intelligibilities of the object are increased as the Intelligence of the perceiving Self is unfolded. So in the stone, when it falls, there is a force of attraction, which is the same force which aids to hold the systems of worlds in order; so in the attractions of the infinitesimal particles which aid in building up vegetal and animal structures; so in the instinctive and unconscious love

of the animal dam to its young; so in the love of the
human mother there is an attraction to the child which
becomes conscious, with various modifications derived
from the æsthetic and the moral culture of life, but the
simple attraction is the root-force around which they
racemate and cluster; so in the autonomic forces, by
which the vegetable grows, there is a cunning wisdom
hid, by which it produces and moulds and fashions its
particular autonomy, attracting and carrying from root
to topmost leaf and bloom the plastic elements of growth
and continuous production; so in the animal autonomy,
— and the intelligibilities become more complex and
yet more open as the variety in the organic structures
increases, and the variety of structure increases as the
variety of manifestation of instinctive powers increases,
or, more properly, e converso, the powers inwoven in the
unconscious instincts become more open with an increase
of organization in this ascent, and, finally, in that break
and separation between the afferent nerves by which
sensations are carried in to the Self, and the efferent
nerves by which motion and actuation are carried out
into life, whereby man is enabled to interpose his consciously autopsic and controlling or directive powers, the
Self appears clothed in the majestic power of self-normalation and of unfolding its higher life. I. i. 29; iii. 5,
6, 9, 14, 15. These Intelligibilities, thus inwoven in the
orders of creation, are correlatively differentiate and perfect of their kinds, from the atom upward, and so far
as they are or can be made the subject-objects of intusception they give forth these Intelligibilities in their respective forms. So far as they can be psychologized, in
the method herein unfolded, the intuscepting agent and
the intusceptible object both increasing, the one in the
conscious exercise of these triplicate powers and the

other in the character and open exhibition of these intelligibilities inwoven in the latter, there has been found
and will be further found corresponding correlations of
higher organisms, and, as the ascent is made, it may be,
nay, must be apprehended, that, as the intelligibilities are
more perfectly and openly inwoven in the object, and
the more complete the intelligence inwoven in the Self
in correlation with its consociate powers, higher existences may know each other as a man knows his own
thoughts and loves and actuating power. So God knows
all things, and so is he known to Intelligences by the Intelligibilities he has concreted into his creations.

28. This is the Method. It is analysis empirically
applied by a threefold intusceptiveness. When such
analysis is perfected in whole or in part, when the subject and the object are apprehended in their mutual correlations, and each in its inner correlations, and in their
positions and relations to surrounding things, and have
been conjoined, articulated, and synthesized into system
and form, in the whole or in parts, as far as these processes
have been carried, then so far the subject and the object
are understood — comprehended. In a perfect analysis,
as the process proceeds, the constructive elements and
the constitutive forces are all arranged in their due
proportions and positions according to the transcendental
idea of their correlate and systematic combinations and
adjustabilities, — that is, according to the intelligibilities
inwoven in the whole and each adjusted and adjustable
part. To grasp the Whole, τo Πav, is the task of the
triplicate Self. I. iii. 28–31. This is the order of human
progress in its yearning love and intellectually normalative approach by deeds of intelligent beneficent actuation, ascending up towards that Supreme, whose ordination rules error and the wrong of individuals and of

nations, and resolves them into causes of improvement
and beneficence, yet so as to show that those who live to
the animal become as the animal in the lower appetites,
and those who live as man but more thoroughly inweave
into their lives human passions and affections, and that
those who normalate their lives, and produce the righteousness which is sown in peace of them that make
peace, become depurated from the orgasmic forces of the
soul and ascend in the serenity and gentle firmness of
the spiritual, the diviner life. I. iii. 14, 15.

29. In all cases of self-conscious reflex knowledge
there is the process of Intusception. The Self can only
grasp the symbol in its form or in the ideation of its construction, so far as it can penetrate, infold, permeate, in-
• fecundate, enlife — in some sort, actualize, by this living
reproductive, representatively creative process. The
horse can see the man, but, so far as has been ascertained or is at all philosophically probable, none of his
intelligence as intelligence. It requires a consciously
autopsic Self, having these triplicate elements capable
of being modified and functionalized into the forms of
action, motion, and actuation, intellective faculties, and
wants, appetites, desires, &c., to cognize and comprehend,
the elements of these manifestations in itself and in other
existences or in their coessential efficiencies in Being.
A less intelligence can grasp so much of a higher nature
as to it is made intelligible, and this can be only in virtue
of its own intellectivity ; and so as to the affectional element, and so as to actualizing power. As these become dis-
sociated in man, mania and monomania in various forms
of automacy supervene ; and their dissociation in any
ideation of God will destroy every possible attribute of
an intelligible Deity. And so the Self passes through
the painted chambers of imagery furnished by the

organisms from the finite symbols and by its intuitions
and ideations to their filamentary correlations of system
and order, and ascends to that ultimate morality which
is the pure love of God — and the Proleptic Morality is
again seen but as a ladder resting on the earth and reaching to Heaven, and the " angels of God ascending and
descending thereon."

30. For man there is no moral truth until it becomes
the possession of his spirit, and is inwoven into his
soul for actuation into life. Then he will find or form a
symbol, a creed, a system for the truth, in the genial,
inspiring, yet inexorable processes of the Moral Logic,
I. iii. 28–31 ; and thus it will recreate the Life of the
Truth in his own nature and in others, and society will
be energized with the Thought impelled by his love and
actuated and thus actualized by him into the current
of life. In reaching this clear elevation, the conscious
Self begins at zero with sensation, and passes through the
intusceptions by the Self, in its own spontaneities and
normal processes, to the intuitions and ideations, and
normalates and articulates the whole into a syntactic
system. The whole is synthesized, and thus he sees how
all unite and correlate, and he ascends by his gradual
analytic-synthesis to the Omnipotent, Omniscient, Loving
One, and sees that in Him Creative Power, Intellectivity,
and Love are eternal and coessential coördinations.
I. ii. 19.

31. Such is the analytic-synthetic-redactive process
by Intusception. This combination of analysis, synthesis,
and redaction, separating these facts and re-living these
processes and gaining the intuitions and ideations, and
articulating the underlying forces in their systems of
correlation, necessitates their comprehensive and orderly
manifestation in the unitary and unifying coördinates of

a single Divine Consciousness — an all-comprehensive synthesis of a Personality. And this method and its results give a single consciousness in man — an Unity in a creative, intellective, loving, and — in the philosophic history of man as in the apocalyptic history of faith — an Educative God, in the ascending movement of his Prolepsis.

This is the Law and the Method of an entire and replete Self, in the integrity of its trinal elements and in the perfection of its organisms, which produces Form in its final results and gives a central complexus for the insistence of forces and their harmonious working and manifestation ; and this, in the highest ideation, when all these Forces are united in a harmonious centre or unity of action, is Personality. And so the intusceptive method seizes the Initiate Causations in their insistency and unitive modality. In the completed normalation of the forces of life into thought, love, and act, the modality of the forces are executed. Life is the picture of the inner life, and the highest form of life is Personality. I. i. 40.

BOOK FIRST.

THE TEACHER AND THE LEARNER.

———◆———

CHAPTER FIFTH.

THE TOOLS, THE INSTRUMENTALITIES, AND THEIR USES.

1. " THE philosophic Hermes says, What is conversation between man and man? It is a mutual intercourse of speaking and hearing; to the speaker 't is to
teach; to the hearer 't is to learn. To the speaker 't is to
descend from ideas to words; to the hearer 't is to *ascend*
from words to ideas. If the hearer in this ascent (cannot arrive) at the ideas, then he is said not to understand.
If he ascends to ideas heterogeneous and dissimilar from
the speaker's, then he is said to misunderstand. What
then is that he may be said to understand? That he
should ascend to certain ideas treasured up within himself correspondent and similar to those within the speaker
The same may be said of a writer and reader." Ad-
mitting the correctness of this, under limitations and
only when the speaker and the writer express intuitions,
and have themselves reached the final ideations of the
transcendental system — those things which are always
and under all circumstances true as belonging to the In-

sistent Truth and the Divine Ideas, I. i. 35–37, it will
be seen that this is the method of communication between God and man — but this only as man progresses
from his individual and historical abasement in the movement of the Divine Prolepsis.

2. The inverse action here mentioned, namely, the
ascent from the word, the human symbol to the
thought, ideate, in the intercourse between man and
man, and the ascent from the symbol, word, sign, or
created thing to the *idea* in God the Creator of the
symbol and teacher of the idea, in the living forces inwoven into the symbols of nature and of life, can only
be by Intusception, carrying with it the elements of the
threefold life treasured up in the Self. The Self is a
centre of self-consciousness and a point of intellective,
affective, and actuating deployment, and in the acts of
intusception there is a going forth, as it were, of and by
the Self from its own centre of triplicate activities to,
into, among, around the objects, whether of matter or
of mind, or of objective states of the mind, and infusing,
interfusing, and circumfusing, — inflowing from· and by
the Self whatever is cognized or to be cognized. I. i. 2 ;
iv. 1–4, 28–31. The Self is, as it were, in this aggregate
of its movements a moving presence — a fascicle of aural
forces — of telegraphic presences to be intuscepted into
and around the mould of the symbol and thus to take
its form and gain its life-contents. In the cognition of
external things of sense the self impinges in this manner
on the limits of the symbol, or makes its contacts by
means of the Senses, while it must in like manner enter
within all animate and moving symbols, and catch the
forces at work within or without. And precisely as
these operations are limited or enlarged by or for, yet
always by the Self, in its capacity or exercise of cogni-

tion of quantities and qualities, in the sense-organisms, are its cognitions limited and is thus limited in its cognitions — its knowings, in this direction.

3. As the Self advances and passes from the cognitions of mere external objects, it finds certain sensations within itself produced by the natural actions of different parts of its own autonomy common to all animal natures, as hunger, thirst, &c., and in after-life its erotic desires and other impulsions ; still later in life it becomes self-conscious and consciously self-controlling — self-directive - of its human impulsions to action in the various circumstances and contingencies of life, as connected with both its own animalistic and human correlations to society, arising out of its sociological instincts and its necessities for determinate action, and it moves on and reaches its more or less clear intuitions and ideations in the sympathies — the unitary coherence of its own solidaric forces, — as well as in the sympathies of these forces with the identical forces in the lives of others it is brought into conflict and harmonies in life, and evolves its own nature and the natures of others. In this advance, the momenta in and through all the senses and by their organic means brought into contact with the Self by the joint action and reaction of the two, — these momenta acting on the Self and the Self reacting on them, — the sense-momenta cease to give instruction, but the Self attains to higher cognitions lying behind the sensible images and the sense-impulsions, yet still through these as recorded in effects in and upon its immediate organisms and upon other parts of the system by the connections, the inter-correlations which subsist between them, and by which the different parts of the viscera are affected. I. iii. 5–15. It sees life, force, motion, phenomena of various designate kinds, all around it. It is

conscious of life, force, and motion, in designate kinds
of phenomena within itself. It seizes these in its own
conscious self-actuation, self-restraint of actuation, self-
direction, and in reflective judgments and in the affectional movements. It uses the forces of nature and life,
physical and moral, as it learned, cognized them through
its various senses, external and internal, and in its own
conscious introspection and in its intusception of other
kindred natures, animal and human, and in its progress
it would comprehend these forces and write the laws of
the universe. These it cannot understand until the Self
itself grasps its own fundamental powers in greater or
less clearness, both as a conscious appropriation of these
powers, and as they have actualized the powers of nature
and life, and as they actuate them. The painful experience of the human races by which the Self ascribes so
many of the phenomena of nature to the direct interpositions of gods conceived by itself so much like its own
undeveloped and unnormalated powers, and by which
it makes a god for different manifestations of nature and
weaves in such conceptions passions so much like its
own, is but the presentimental spontaneity of the human
creature grasping-at intelligible forces by which the
orders of nature are animated, or, as it supposes, violated
by interpositions. From these crudities it passes to the
recognition of uniformity in the action and mutations of
nature, and reaches up to the generalization of laws from
on this side of nature, and finally gains the ideation of
law-forces from on the other side of nature and life, but
can only see laws thus-wise, as inwoven into organized
system for the movement of forces. But the Self
can get no law for the action of anything until it, in
some degree or meaning, gets a law for and from itself.
The stone falls and it will crush, but it has no conscious

power, but the power, force, by which it falls is, at length, called gravitation; a name is given, but an essential force is not understood. I. iv. 3, 4. But in process of time it is seen that gravitation has a wise law inwoven and concreted in it — that it is a law-force executing itself. It is a force acting blindly and constantly, but, by its wise correlations in quantities, and relations of time and space, it moved the suns, planets, and star-systems, and the intelligibility which the Self gains from its correlated action with the other two forces is but the exact correspondent, so far, of the Intelligence which made the correlations. The fact of gravitation has the new element of intelligibility discovered in it by its correlations with the other forces, and the other forces must have been cognized in their oppositions and redactive control before the laws of action of either, in this their combination, could be understood, and the substances of matter are supplied to give a content to these abstract laws — ideas. The horse has no moral sense, and yet may kick and kill. A force is becoming open and intelligible, — it is a centrifugal force, — but whatever distinguishing intelligibility is in this new form of force is perceptible by virtue of the conscious intelligence of the observer, which uses the same force consciously, from which he intuscepts the *exerted* force of the horse. The tigress or other animal dam has an intense love of offspring attracting it to its young and its place of habitation, which ceases after a few months, and revives with new offspring, but never reaches a moral nature. In man these become independent, autopsic, and moral; and from this position, in virtue of his conscious possession of actuating power, attracting affinities, and intellective direction and control of these powers, by his own autopsic power he intuscepts the forms and forces given to him by his senses from

nature and life, and thus, in and by means of his own
conscious intelligibilities, he comprehends the intelligibilities inwoven and implexed in nature and life. Beyond
this there is no comprehension — supposable even for
Deity. His Intelligibilities descend into creation.

4. Man is surrounded and constantly conversant with
representative symbols, as things, things made, artistic
executions, language, acts, movements. ; Deity when he
creates, must create intelligibly in form, in forms. I. i.
23, 24–31, 35–37. Precisely as the Self embraces in
its consciousness the symbol and gets *back* of it — ascends beyond the form, the symbol, the thing, the language, act, or movement, and comprehends the movement-
forces, it understands the powers — forces producing the
symbol and giving it its movement-forces, whether in the
movements of nature and of life, the conscious conduct
of men or the creations of Deity. Thus he ascends,
by these acts has ascended from the sensible to the supersensible intelligibilities in man — to the suprasensible
powers as forces in God, yet always given *through* the
Sensible. I. ii. 13, 16, 18 ; iii. 9, 11–15, 20 ; iv. 3, 4.

5. A symbol may, and perhaps constantly does,
represent several ideas, namely, the mere constructive,
working intent, the forms of its general use, a special
end or purpose, and this as subordinate to some intermediary end further removed towards some ultimate end,
and its particular and appropriate form, and its ornamen-
'tation, and all with correlations to other things as parts
of a systematic whole. Thus language may be rigidly
intellective, yet may be conjoined with demonstrative
power from the anger or wrath, or from the suasive
gentleness and intoning mercy of the affections ; and it
may be ornamented. And in all it works into form. In
the watch, most men understand *its* final end — cause, as

an instrument to measure time, while few understand its
exact combinations, and how by a variety of different
forms in the subordinate parts, and how by a variety of
different modes of construction — in all combining centrifugal and centripetal forces subordinated in redactive
forms and thus all intellectively correlated to the given
end, it produces that measurement of time, while there
may be various opinions and tastes as to its ornamen-
tation. But to ascend higher: most men have some
sense of duty, and to some extent feel the obligations
of *the* Right, but can neither comprehend the complex
arrangements of the dependencies in life by which
questions of right and wrong are constantly presented
to their election in the ever-changing vicissitudes of individual life and the conflicts of nations, compelling *a*
choice but not *the* choice; nor has any one propounded
a satisfactory principle, element, or combined action of
elements in the Self on which to base any acceptable
philosophy of this Sense of Responsibility, and its appearance and growth in the individual and in the relative
conditions of the tribes and nations of men. II. vi.

6. In the process of intusception it is evident that the
Self must first apprehend the symbol as thing, language,
act, movement. The cognition must be first of these,
and this in some of the ways in sensation — as in some
way registrated through the organism. I. iv. 24. Having the *factum*, the symbol, how attain the complete
ideate of it, if made, fashioned by man? How get
back to the intellective form from which it was so made?
Clearly, by going into the processes of making, as
these processes are given or may be attained by the
self's own powers. I. iv. 28–31. In this way the mode
of making, and so the motive-end for making, in the
many various motives for making, are attained. How

attain the idea of the symbol if it is a thing made —
created by Deity? The process is the same, and is that
instituted herein. But there is a class of Truths beside
the Mathematical Intuitions, as Ideas — the forms of
things and their correlations in the Divine system, and the
Moral Ideations. The first, mathematical truths, are pure
intellective *forms* into which quantities may be adjusted,
I. iii. 28, 29 ; iv. 25, and are therefore seen intellectively
— intuitively ; the second are those forms preparatory to
or directly appropriate to the constitution of the vegetal
and animal life as leading to and subordinate to and for
the uses of the autopsic life, and therefore embrace other
elements than are presented as concreted in the pure intellective intuitions and actualized in nature ; while in
the third, both of the former are combined with a more
open and expressive use of those elements which confer
on the autopsic Self Loving, Thinking, and Actuation.
The first are seen through the simple mathematical symbols; the second, in forms of actual, positive creatures
and things in nature and life ; and the third, in the harmonies and conflicts of responsible beings, yet all in and
from the self. I. i. 23, 24, 28, 31, 38–40 ; iv. 3, 4, 18, 19,
22, 28–31. In this view, and only in this, the aphorism
of the Scholastics is true — there is nothing in the intellect which was not first in the Sense: *nihil est in
intellectu quod prius non fuerit in sensu.*

7. As a primary act of intusception let the first step
suggested be, as it is in life, a simple apprehension of a
material symbol. Here it must be of the symbol itself,
and thus be by sensations derived through and by means
of the Senses, each in its specific action. The blind can
have no perception of colors, and therefore to the blind
the symbol is colorless ; the deaf cannot hear, and to him
it is without sound ; and if presented to paralyzed nerves

of feeling, the Self will have no percepts of hardness, &c. ; and so on through all the senses, external and internal. No sensations, no perceptions ; and no perceptions, no quantity, nor quality. No sensate, no imaginate, no thought, no judgment, no comparison of past moment with the present, no intuition, no ideation. Life is a blank.

8. So Sensations are the first steps towards cognitions. No sensation, no cognition ; nothing to intuscept, no *mean* for intusception. Sensations are impressions upon and modifications of some nervous tissue or filament capable of conducting such impressive force to the common centre of perception — the Self. I. iii. 7–15. To perceive there is a *nisus* — an .attention, attending of the Self, going to or into the filamentary thread of nerves, or to the convolute of the brain in which it, each, discriminately terminates, and there receiving the impression sent in. It must go, as in § 2, *ante,* to the convolute and attend to the sensations and possess itself of the representative facts which it brings. So in each case to each discreet organism, whether of sight, hearing, or hunger, or animalistic desire. It is true these may be colored by differences in the organizations of different persons, as one is incapable of seeing certain colors as others see them ; one prefers the sweet, another the sour ; and in looking at one object, different persons will more readily be attracted to one part or quality, and others to others. Yet it is from these objects and through such organisms the Self makes its cognitions. The speciality of the afferent nerves, thus various in their functions of intelegraphing in to the Self, and thus diversified in their modified action, but gives the obverse side to all that organization expedient and necessary for the direct action from the self, outwardly and among the internal organisms, as also of the repercussive action or interaction

13

from one animalistic or soul organism to another.   So in
like manner the speciality of the animalistic and soul-
organisms are given in their diversifications and differ
ences.

9. When the Self takes cognizance of these impressions or modifications of the nerves, it is perception, — its
observations here are percepts.   In the fact of attending
to and perceiving the sensations they are registered in
the Self or in its correlate organism, and so there is an
alterability and an alteration within.   That which was
before without a'particular perception has become so far
altered that the perception is impressed, — in some way
inwrought into it.   There is always, in such processes, a
Registration.   These registrations are so engraved that
they can be re-produced, and thus read by the Self.   They
are then Imaginates.   § 26, *post*.   It will be seen clearly
hereafter, II. ii., that there are Registrations of impressions actually impressed and engraved from without inwardly, and this whether from the internal or external
senses.   There is always a registration on some definite
and respondent portion of the nervous system and on
special viscera.   It will be further seen, and is affirmed,
that there are actions or spontaneous movements lying
deeper in towards the Self, originating in the psychical
organisms, such as the desires natural and native to the
human soul, love of acquisition, domestic emotions, &c., &c.,
which produce impulsions of various kinds to action and
demonstrate in outward conduct.   These are also engraved,
and sometimes deeply, in their action on the system, as is
seen and felt in those emotions which so deeply *effect* the
different portions of the system, and leave their permanent
effects on the countenance in their engravings of general
character, and in the pleasures or retributions of memory.
I. iii. 9 ; i. 1.   These distinctive effects are manifested

from and by their special orgasmic forces, and as they incite to this or that line of conduct. These are unnamed, and may be fitly termed Psycognoses or Psytations. In the animal and the man they are seen to repercuss through the nervous organizations from one to the other. I. iii. 5-15. They, too, are seen to have their correlations and adjustabilities in and among the external objects of nature and life. I. i. 29, 30, 33 ; ii. 6, 7, 18 ; iv. 11, 23, 27. In the further unfolding of the solidaric Self from its envelopment in the somatic and the soul-organism, it is seen that the Self, in its solidaric simplicity within, sends down its conscious powers to curb, control, and direct the animalistic impulsions, and that in like manner it moulds and modifies and subjugates these psychical movements, and that it plays one orgasmic force on to and against another, and that it, too, can, within certain limits, *notate* its powers, as positive forces, in the different parts of the system, and thus modifies the organization. I. iii. 15. For where the organism or the viscera are affected by any passion or affection, I. iv. 17, and these can be controlled by the conscious Self, there the power of the Self, by direct agency or by withholding and controlling the forces which produce these effects, must extend.

10. When the Self would cognize through any one of the sense-organisms, it must give attention to the sensations produced through the nerves of the special organism affected by its appropriate external object or quality of the object, or its internal orgasm, and which is thus made the object of observation. § 7. These sensations, so affecting each special organism, must be communicated to the Self at its common centre in the brain for receiving information from and through its various organisms, for if the organ at its proper point for receiving its special

impression is destroyed, or if not there injured, but the communicating, the afferent nerve is divided at some intermediate point, Perception — cognition does not take place. So in the various movements of life, on presentation through the Senses of the appropriate objects, various passions and affections in the soul-organisms are excited, — " the sight of means to do ill-deeds makes ill-deeds done." So through all the Passions and Affections. At this conjuncture it is seen that the Self comes down, as it were, with its powers of autopsic interposition, to control and subjugate and play off the better natures against the worse, or to normalate the whole into some system of life, animalistic, human, or spiritual ; and in the walks of common life it may daily be seen, and constantly in the wards of the Lunatic Asylum is seen, how various portions of the organisms in their abnormal orgasmic conditions become automatic, and that their special forces or manias govern and overrule the proper Self. There is but little proper Sanity. These effects registrate themselves in various parts of the viscera in legible characteristics. So it is with the conscious Notations from the Self ; and the more frequently these are made, the more easily they are performed, until the whole organisms, functionalized with these higher powers, exhibit the higher life within in the uniformity of their daily action and their characterizing effects on the whole organization. They are thus seen as registrating forces which leave their characteristics deeply and indelibly and intelligibly written in the living statue of the Self. II. iii. iv. v. vi. vii. viii.

11. It will be seen more fully hereafter, II. ii. iii. iv., that in the lowest animal organizations, their systems are so constructed that the afferent nerves and the nerves of motion through which these animals act for the sustentation of life and the continuance of their species, are essen-

tially internuncial; that is, impressions made upon the afferent fibres excite respondent or reflex movements in the efferent or motor nerves, " without any necessary intervention of consciousness;" the telegraphic circle is completed by the cause which commenced the action in the afferent nerve, and it is consummated in virtue of the impelling force thus continuously communicated. There is no *break* in the circle. I. i. 30. It is here necessarily perceived that the impression is made on the afferent nerves, and that the reflex action of the motor nerve is but the original causative force in continuous action, and that therefore the lower orders of animals especially and conspicuously perform the functions of life without self-consciousness, and solely in virtue of the impulsion given to the afferent nerves being continued to the efferent motor nerves and producing their appropriate respondent action, *e. g.* just as anything which touches the arms of the Polype is entrapped by them and drawn into its stomach. I. iii. 11, 12, 15. It looks and acts like cause and effect in the physical world; and when the same acts or phenomena are seen in the beautiful flower, — Venus's-flytrap, the *Dionæa muscipula,* — it is so called. So is the action of confirmed habits of every kind, to a certain degree. So are the manias and the monomanias and exacerbated fanaticisms. These facts may be seen in the offices of the nervous system as they are becoming understood and appreciated by the physiologists. " Every collection of gray matter, whatever be its situation or relative size in the nervous system, is called a *ganglion,* or nervous centre. Its function is to receive nervous impressions, conveyed to it by the nervous filaments, and to send out by them impulses which are to be transmitted to distant organs." " One set of these fibres run from the sensitive surfaces to the ganglion, and convey the nervous impression *inward*. These are called sensitive fibres" (the Afferent nerves of Carpenter). " The other set run from the ganglion to the muscles, and carry the nervous impression *outward*. These are called the *motor* fibres " (the Efferent of Carpenter). " We have already stated that the proper function of the nervous system is to enable a stimulus, acting upon one organ, to produce motion or excitement in another." " This is called the ' reflex' of the nervous system ; because the stimulus is first sent inward by the nerves of the integument, and then returned or reflected back from the *ganglion* upon the *muscles*" (the Internuncial communication of Carpenter). " It must be recollected that this action does not necessarily indicate any sensation or volition, nor even any consciousness on the part of the animal. The function of the gray matter is simply to receive the impulse conveyed to it, and to reflect or send back another ; and this may be accomplished altogether involuntarily, and without the existence of any conscious perception." — Dalton's *H. Phys.*, 365, 367, 366. Thus it is seen how the vegetal life acts like physical cause and effect ; how the same kinds of causations and effects are inwoven into the unconscious animal orders ; and how the same are in-woven in the manifold organizations of the higher animal orders and in man, and there inwoven with instincts and autopsic powers. And man soon becomes conscious that certain forces originate within, without any consciously originating power or purpose within — the psytations ; not only so, but that they manifest their action in opposition to and in spite of his self-determining purpose, and that they observantly manifest themselves in physical alterations of the organization ; — he cannot blush when he wishes to, nor feel fear, or awe, nor have the feeling of charity. Some of these various effects seem to come

without cause, yet are always referrible to cause, and in
all of them the Self is capable of originating momenta
for their control and cultivation. Civilization teems with
the effects of these normalative causations, while evangelization is ever and forever " a burning and a shining
light." In this normalative conduct he becomes self-
consciously conscious of originating powers — forces, to act
from within outwardly, as of conscious force to move the
arm, and a problem in thinking is consciously arranged
in its order and terms, and delivered to the tongue to be
uttered, or to the hand to be written, making it the tongue
of the written thought ; or the feeling of charity may be
submitted to the intellectivity, as to the propriety at the
time and in the object and for the best means of accom-
plishing the end, and then delivered over to the actuation ;
— so man makes the watch, the engine, the book. And
in all these processes it will be seen, by those who are
capable of making the introspective examination of the
Self, that it always acts from the impulsions of the
animalistic forces, the human orgasms, or from a higher
life which either subordinates or modifies all these.
Each step upwards makes the processes of life more
transparent and perspicuous. In thus passing through
these classes of phenomena it is seen that there are impulsions to action from the lowest animal instincts incorporated into and made part of the human economy, and
that there are impulsions more distinctly characteristic of
the human nature ; and that in both instances it is the
economy of these impulsions to execute themselves spontaneously, — as in the flytrap or the polype, as in the
animal instincts, as in the daily natural impulses of all
human conduct, and as in the fixed habitudes of life.
The impulsion communicated from its centre of origin
tends to produce the respondent acts of the other parts of

the system necessary to secure the gratification of the impulse. Hence the folly, the viciousness, if not the guilt of adding coarse habitudes of life to those we already possess. The human characteristics are inwoven and concreted in organisms in a manner similar, if not identical, with that of the animal instincts in their different specialities; only in man they are of higher endowment, and more varied in their characterizing powers, both in the original function of the respective differentiate orgasms and consequent organisms, and in their combinations and inter-correlations between them and each other and the Proper Self, by which their functionalized powers may be improved or increased, or diminished or modified by the conscious control and use by this Proper Self — to beastliness, crime, or holiness. It is, here, the conscious Autopsy in man, with its self-generating powers by which it resists and breaks up the circles of these animalistic and human impulsions, or conditions one with modifications from the other forces brought into action, and thus directs, controls, and modifies its life. In thus breaking up the circle of impulsions either from the animalistic or the psychical nature, it is manifest that a force or forces are necessary for intervention and discriminate direction of action and restraint of action in these forces. Thus all the forces are again seen to be causative and interactive, and as such they too registrate their effects in the organization. I. i. 1 ; iii. 5–15. The means and mode of action in the animalistic and human conditions are by orgasmic forces originating in ganglionic centres and communicated from one part of the system to another, and, in self-conscious man, to the Proper Self by the nervous tissues. This is openly conspicuous in the erotic passion, the seat of which is so far from the sensorium, the point of radiating power from this proper Self. And this ex-

emplifies the origination and the modes of intercurrence
of all the animalistic and human orgasms, as it shows
that the reaction back upon them by the consciously
• ruling power of the Self is by intercurrent nervous fibres
placed or which, in the pure culture of life, may be
brought under the control and use of the Spiritual Self.

12. This analysis and synthesis give the outward
impinging forces acting on the eye, the ear, the taste, the
touch, the smell, and the registration of their effects inwardly to and for the Self. They also give the animalistic
impulsions from the stomach, the venery, &c., and their
like registration. So in like manner those impulsions
peculiar to man, either in their distinct natures as wholly
human, or as of higher organic constructure than those in
the animal, yet of the same kind, have their organs of
origin, yet by supply of forces from the general sources of
assimilation in and through the organization of the system;
and they have their nervous connections, by which they
intercorrelate with each other and registrate their demonstrative effects in various viscera, — shame on the cheek,
mercy on the bowels, fear on the heart, awe on the hair,
&c. Each class of sensitive and sensational forces moves
along their appropriate afferent nerves to the Central
Point where the self-generating autopsy in man can consciously move towards and attend to the sensations thus
brought, and discriminate the objects and sources from
which they come, and distinguish them and the sensations
thus brought into their various classes, referring each
class of qualities to its appropriate subject and each to
the particular organism through which it was intele-
graphed to the Self. It distinguishes many differences
in the sensations of the same organ, as of colors in sight,
savors in taste, sounds in the ear; and, in later and more
advanced observation, discriminates in the internal cau-

sations, producing their various effects as before brought
into view, and in their differences of kinds and intensities. It thus gets, also, the direct and general effects of
various stimuli and medicinal agents in increasing or re-
tarding the general action of the system and the specific
action on various ganglionic centres and on the specific
animalistic and human orgasms, still showing the
affinities and kinhood of the forces and yet their differences, for in all these effects there is a yielding, a setting
free of something in the organisms affected, and there
is something gained from the agent used — there is a
composition and resolution of forces.

13. In these discriminating cognitions by the Self is
found the Self and the Not-Self. The fœtal child may
have sensations, but they are purely subjective. Carp.,
H. P. § 591. After birth it learns to discriminate
objects by these differences of sensations conveyed from
the objects to this inner Self. In the same manner, all
the cognitions by the Self, of its animalistic and human
impulsions, are percepted and discriminated. Perceiving,
cognizing, that objects produce their certain, constant,
diverse effects in kind, it discriminates them in their
differences in kinds and degrees, in and through the articulate organs correlated to convey them, — heat, color,
smoothness, &c. § 7. So in the discriminations of food,
and so through the orders of the internal senses. In
like manner it catches, in "the wear and tear" of life,
the effects of the Passions and the Affections of the Soul
by the sensations produced in their various correspondent
viscera. In a further advance it finds its whole Affectional nature as a pure, serener enjoyment in the Inner
Self in the contemplation of purity, order, and justice,
and it finds its intellectivity no longer as the respondent,
the repercussive agent of the animalistic or the human

orgasms, but it becomes self-conscious of its presidency over all, guiding, directing, and elevating the sanctified Actuation, in means of comprehensive beneficence, to deeds of gentleness and serene Love. These discriminations are the processes of long years and of advanced position, however attained, in the unfolding movement of the Prolepsis. In this movement forward there is, in the very nature of these causes and effects, a constant action and reaction of the Subjective and the Objective, and their forces interweave imperceptibly, yet always opening up into wide diversifications. The Self, all along the line, makes its cognitions through the senses, external and internal, the animalistic and the human desires and impulsions and their repercussions and its own independent notations — all acting and reacting on each other and furnishing to the Self the full complement of cognitions and growth, and presenting the symbols for Intuition and Ideation. I. iv. 3, 4, 17, 18–20, 24, 26–28.

14. The Central Self acting through its notations is the centre of self-generating, or it may be of original yet self-directing forces. Life, so far as cognizable, is the action and reaction of forces. These forces are seen as they pass from mere physical nature through the vegetal, the animal, and the human orgasmic forces to autopsic life to disenvelop from the coarser physical nature until they appear as open and conscious forces, in their triplicity, and come as it were from above, out of the Self, and react on nature and on the animal and psychic orgasms. So when we go back to a creative Deity. § 11, *ante*. In the laws of forces, force can only be modified by force; but it is only at this height that the controlling, modifying force of the Self becomes self-consciously autopsic. This gives an antagonism of contrapellent forces, namely, the forces in external nature and in the internal animalistic

and psychical organisms acting towards and upon this
centre, and the central Self acting on these forces, as in
the harmonies of life which supervene on the supremacy
of the higher autopsic life — harmonies of adjustable correlations. And as these unfold we can, like a thread of
silver light, see Order track its way through the dark and
bloody centuries, and

> " through plots and counterplots —
> Through gain and loss — through glory and disgrace —
> Along the plains where passionate Discord rears
> Eternal Babel — still the *holy* stream
> Of moral life roll on."

15. Form now the ideation of a first distinct act of
creating. The Spiritual Forces have not gone over into
Actualization — into creative manifestation; all space
is empty; God is omnipresent in his spiritual, essential
forces; the act of creating as phenomenalized in matter
in this planet is of a point or aggregate of points, whether
of *atoms*, as of the Greek Stoics and Naturalists as following Democritus, or the *homœomeriai* of Anaxagoras, or
the *psigmata* of Heraclitus, or the *monads* of Pythagoras
or Leibnitz, or the *molecules* of those who seek a more
general expression of these *infinitesimals* of which all
bodies are composed. I. ii. 12. And these are posited
in space. The act of creating may be, in the facts of
nature must be, ideated as an effect of forces, and as
such a combination of forces and not otherwise. This
combination of forces, in one point of space, may be conceived as enlarging by the point pressing through itself
— as of the mathematical definition of a line — a point
in flux. This would give the ideation of the production
or creation of a common or homogeneous mass being extruded and, without differentiation, objectified into space
and without specific form. To give it specific form

there must be other forces controlling it into specific form, and to give it differentiation there must be the conception of these forces in differentiate combinations. To give it specific and appropriate form and an on-going of adjustable correlations there must be an intelligent directive force countervailing and inweaving and concreting the forces. This gives the ideation of objectivfacient, sustaining, shaping, individualizing, and actuating forces. The increase in number, volume, and adjustable correlations of the objectivities produced increases the volume and the triplicate character of the evidence for the Powers of the Creative Forces. I. iv. 27. And thus it is that the question is constantly presented whether is right, the generally received doctrine that matter is an eternal substance, and that as such it is a vehicle of forces, or the hypothesis of Boscovich and Faraday, that forces are the only existences, and that matter, in its various presentations to man, is but the composition and resolution of forces. But matter does not appear as a homogeneous mass, either in the larger accretion of bodies or in the infinitesimals of the combinations which form the Infusoria. All known matter is atomical, and in the infinitesimal molecules is differentiate, as in the shells of the Infusoria some are siliceous and others are of lime. To reach the ideation of diverse creative forces and of these stabilitating certain things and providing for the phenomenalizations of all, the Self must perceive the powers — forces of Deity as ubiquitous, and many points of matter or flux of forces will then be intuscepted as appearing at once or in many points of space, and forming matter and natural forces, under the unitary control of the Autopsic Being. These are things and natural forces — the created. These atoms are seen, at once, as held together, and yet as kept separate, in a certain sense, by the very forces necessary to form the planetary systems, — the Dynamic forces embrace them at once. They are seen as the constitutive forces of the things created and as their governing and controlling powers. The multiplicity of material elements, and of forces acting in combination and connection with them, necessitate the conception, the ideation of a power, force for sending forth, objectifying, setting over in objectivity from the Creator ; I. i. 8–10, 17–29 ; ii. 2–6 ; iv. ; force for aggregating, collecting, combining, cohering as of specific yet differentiate attraction, — a retractile, static, dynamic force of attraction for drawing together correlated created points, atoms, molecules, or monads ; a repellent, centrifugal, antagonizing force to maintain the separateness and independency of their separate existences ; and a designing, arranging, wisely controlling, correlating, superintending, form-giving, directive, redactive force. The thing produced under the operation of these forces is the symbol from the Redactive Force, yet, in the production of its organisms, orgasms, functions, purpose and gratifications, by the inweavement and concretion of all the forces. II. yiii. Coördinations from on the other side of nature and life are correlations in nature and life. This is the creative idea ; the intellective and the moral momenta by and for which things, *facta*, were created. It is the transcendental system — the deific idea for the creative act and the teleologic *end* of the Creator. Form enters at once as the law and a law-force of the movement, and throughout the whole the End is in the Beginning. It is the divine idea, or (in the incompetency of human language) *combination* of Divine Ideas which preceded the creative production of the Cosmos — the Universal Symbol — and all of the subsidiary and subordinate symbols. These combinations of all the symbols may be expressed in human language from the necessarily analytic standpoint of the human mind as a separation of Ideas from the eternal omniscience of God and positing them in existences in the order of his prolepsis, thus introducing as it were time into eternity and place in space. I. i. 7, 37, 41–43. These are the ante-types and the movement of the forces into creation, and the orderly on-going of nature and life to the Final Cause. I. i. 15. This is the descent of God the Teacher to Man the Learner. The Prolepsis begins.

16. Ideates are those pictures — mental views which the Self forms, intuscepts of the transcendental ideas, creative actuation, and deific impulse — Love, the Divine Orma. These individual ideations will differ as the solidaric Spirit is disenveloped from the coarser forms of the animalistic and human orgasmic forces and wields its actuous intellective and moral affective powers, and as these are normal-ated in a lower or higher cultus. Properly, Ideation is the product of the Intellectivity as it views the transcendental Omniscience *and* its coördinate Forces in its own clear, dry light of cognition — the *lumen siccus intellectus.* Yet it must always be borne in mind that the intellectivity, acting in and of itself, is but a rationalism, and will forever be productive of error and an insufficient and incomplete system by the omission of important elements in the Created, and consequently in the Creator — the Initiate Causations. I. iv. 2–18, 30, 31. The Self, when it examines, and aspires to grasp Being through the dry, hard light of the Intellectivity alone as solely respondent to its own intellectivity, will to the end catch only one of the Three coördinations of Being — all of which must be grasped to gain the sources of all powers and their correlations in Existences. Yet to attain, obtain these cognitions of the transcendental order of existences before they were inwrought and actualized and

made objective into creation, the Self, as the clear cognizing agent, must intuscept them with its own entire
complement of spiritual forces, but examine them clearly
and, it may be, coldly in the dry light of its mere intel-
leçtivity ; and it must move with the life and vigor of an
undying Love to the heights and depths of creation to
behold their vast unfolding Actualizations. So, while
Ideates are those mental views *completed* in the dry light
of the mere intellective cognition, they are composed of
the actuous, the intellective, and affective forces in their
primary normal elements in the self, and in their correlative activities. It is there seen that precisely as each self
is capable of analyzing back to the foundational movement-forces and recombining them into a syntactic system, so is its actual ability to ascend and understand the
Teacher — God.

17. This gives created points, in their respective forms,
posited in time and space, II. ii. 16, 17, for the production of symbols, and it gives the objectiv-facient production
of symbols in their own inner correlations and outer correlations to nature and life, as it also gives the cognition
of the autopsic self in the fulness of its intusceptive
powers. Then in the production of a symbol, as of a
planet or system, the points, molecules, will be ideated
as appearing multitudinously, but not necessarily succes-
sively. And these points of matter, to be capable of
formation into further and other symbols, cannot be otherwise ideated (it is a necessity of thought, as it is the actual
correlation of existence) than as endowed with internal
adjustments for organic structures, and as possessing intimate and specific correlations of attracting adjustabilities
for the shaping and form-giving autonomies, and of repelling antagonisms to secure the separation and identification of specific symbols, and for the conservation of

identities and the production of secondary causations. This gives the root-forces for the affinitive correlations and the disjunctive contrapellences, (only another form of correlations,) for the production and working efficiencies, and, in the redactive impressment of form, for the production and working efficiencies and preservation of the various organic parts of the symbol, and the independence and autonomy of the symbols themselves. Where there are many symbols — creatures — produced, which must, in the various economies of a world like this, have distinct and separate existences, there must be, not only the correlations to bring together and to unite the homogeneous and the associative elements to make each distinct organism of its kind, but there must be contrapellences, forces of separation, powers of projectile divulsion to separate, to keep separate and maintain the individuality, the differentiation of organs, individuals, species, families, departments, and kingdoms of existences, and especially of species in the vegetal and the animal, seeing that they are constructed out of the same forms of matter and plastic forces, weaving almost identical tissues in both, I. iii. 5–15, and approaching so near to confusion in the similarity of many kinds, and in animals permitting but not inviting hybridization. This gives correlations of harmonies, the unifying, the attractive force ; and it gives contrapellences, the separating, the repellent, the diremptive, projective force ; and it gives the directive, intellective, redactive force. II. viii. 1–8. Thus again analysis, synthesis, and redaction appear not only as the rule of the method, but as the very forces of the method, and are found in the actual processes of the method in intuscepting nature and life, and as the intusception enlarges they unfold in the constant presence of these forces in their movements in the cosmos.

14

18. After distinguishing, in the progress of life, between
subjective and objective sensations, not as a sharp and
consciously instituted process for the purpose of analyzing
them, but as a simple fact of life, the first step in knowledge is the apprehension of the symbol ; and without
symbols so apprehended and intuscepted, the Self can
have no ideate of the transcendental forces which are the
causes of their production and the working efficiencies of
their diverse natures. It catches these forces first as
physical forces ; projectility, attraction, and symbolic form
is present to it everywhere in nature ; when it analyzes
its own passions, affections, and intellectivity, it finds them
uniformly consociated with these forces, — the passions
project, the affections attract, and the intellectivity gives
forms and methods. I. iii. 9–15. Man is so constituted that he must proceed in his apprehension of symbols by this analytic synthetic redactive process ; and
without symbols so apprehended and intuscepted, the Self
can have no ideate of the transcendental forces which are
the causes of their production and the working efficiencies
of these natures so inwoven in the symbols, — I. iv. 16,
17, 28 ; ante, § 6, — that is, it cannot in any degree reach
the why, or what motive-essence as causative-end in the
Maker, for which the symbol was created, the how, or
the causative-forms and method of creating, nor the objectiv-faciency, the setting over in objective immanence
of the creation. Man is so constituted that he must proceed in his apprehension of symbols by a few and limited
number of points at a time, in lines over surfaces, and
in the cognition of qualities by each distinct quality. A
patient and acute psychologist has said, — Hickok's *Men.*
Sci., 117, — " If I would possess any pure diagram in simple mental space, I must in my own intellectual agency
construct it ; *it will not somehow come into the mind itself.*

I can have no pure mathematical line except as in my intellectual agency I assume some point, and produce it through directly contiguous points, conjoining all into one form, and thus I draw the line., Thus of all *pure* figures, simple or complicated." I. iii. 29. Now when the slow processes of infancy are observed in catching knowledge of objects, and if we attend to the *exact* outlining of objects or figures by ourselves or the artist, it will be seen that the process consists in taking a point and producing it. through directly contiguous points, until the exact figure is obtained. If the intellective agent must proceed by points produced into lines to construct the pure figure, much more must the Self in its cognition of objects proceed by points to catch the combinations of infinitesimals in their aggregations and qualities. This clearly will not come somehow of itself into the mind. So the Self proceeds by a very few and limited number of points at a time in its line or process of perceptions through the intervention of sensations. The Self proceeds by lines and surfaces in quantities, and by special forms of sensations in qualities. The Self proceeds and puts these together by its continuous colligation of quantities and qualities, and makes the symbol in its own inner cognitions, thus gathered from external nature ; and when the process is completed, the symbol is apprehended in form, quantities in its parts, and its qualities, to the extent that these operations have been carried through the organisms which have been vouchsafed, and as they have been deteriorated by abuse or improved by genial culture. The Self perceives, and as it perceives, step by step in these processes, it cognizes the sensations brought to it from one part or quality of the symbol and then from another, and thus collects part by part in form, color, taste, smell, &c., &c., — quantity by quantity and quality

by quality as effects from their causes; and retaining in
the memory each part and quality, so ascertained, the
distinct parts and qualities are correlated and synthesized,
by this attractive holding together of the registrating
memory, into the whole of the symbol. *Ante*, §§ 1–8 ; I. i.
38–40. Now it is evident that the Self in these processes,
as it passes from each single perception to each subsequent
perception, loses the prior perception as an actual continuing perception ; yet they are posited in the memory
by some means of registration from the external world,
as also from the orgasmic impulsions ; and the notations
from the Self outwardly are also registrated, for they act
coincidently and are coincidently reproduced. When
reproduced in after-times, they are in the memory. It
is the remembering Self going into the Registrations by
its own reproducing intusceptiveness, and reënlifing them
and bringing them up into reflex observation. I. i. 1–4,
38–40 ; ii. 13, 16 ; iii. 16–19. When this cannot be done,
they cannot be voluntarily reproduced ; yet in virtue of
their associative and associated connections in the Self,
through the impressment made and in virtue of the
elemental affinities of the underlying agencies at work,
they may be associatively reproduced — association of
ideates. This involuntary reproduction, coming when
least expected, coming when the direct intent of the
Self is turned away, shows the independence of the organic functions of the soul-organisms concerned in the
operation, and their inosculation with the Self. The Self,
acting consciously in these processes, reaches the creative
ideations, when it can transcend time and space and see
the Creative Forces going or as they have gone into
actualization — into objectiv-faciency of stabilitations and
moving forces.

19. The symbols of nature constitute an alphabet.

These letters have their syllabification, their words, and their complete sentence, and they point along the returning lines of the radii of their circle to a consistent and harmonizing centre of creative Objectifying Force, Intellectivity, and, as it will be fully seen, of Love, for their fundamental origin and movement. The processes of educating the Self, in the tribal and historical movements of mankind, and in the accuracy of individual culture in self-normalation to the rapid perception of sensations and their conscious registration, is the slow work of infancy in tribes and individuals and in the more matured labors of advanced progress. These facts become more clearly conscious and sharply defined in those higher regions of spiritual knowledge where men escape from the cant of professional life, and reach the practical, if not the metaphysical analysis, that the Actuosity is the executive power of goodness or sin, and seems to work almost independently in the habitudes of the animalistic or the human uses to which it has been applied through continued indulgences or applications; and that, in all new modes of actuation, the intellectivity devises new forms and means and processes by which the actuation is changed and altered to meet the exigencies of a new motive-end presented for action, and the Self by a new direction of action, from a new-renewed knowledge, ascends to a higher life of actuation, or descends to lower and more inveterate indulgences ; and in either case the object for which, the motive-end by which the Self is attracted, is found and only can be found in a love seeking its gratification — in the love of goodness or vice or sinfulness. Thus the Self will have a clear cognition of its own image, of its own intrinsic movement-forces as being guided and controlled by its animalistic and human orgasms, or as controlling these by its spiritual powers to a

higher end of action. In the suggestive highest exaltation of itself it will recognize its supertendencies to the
" height above all heights," and its likeness to Him who
created, and as Creator inwove and integrated into the
created that which was consistent and consimilar to his
own essential powers, his own coördinated, causative, and,
as such, creative forces, yet with a limited independency
on which to hinge the moral responsibilities of life and
the movement of his prolepsis.

20. The processes of the Self, it is becoming apparent,
are of exact analysis, synthesis, and redaction, yet throughout of a triplicate intusceptiveness to catch the movement
of life and of history. When the components of existence
are ascertained, and the specific components of each
symbol are recomposed in the intusceptive reconstruction
of the symbol, and the positions of the symbols are distributed to their proper and correlative and relative syntax
in the whole of the system of the world, and thus their
places in the history of the movements are found, the
Self has reached the idea of creation in its transcendental
purity and the ordination of the prolepsis. This is in
virtue of its intusceptive processes in going into and
seeing whatever is intellective in the Intellectivity, whatever is Actuous and Objectiv-facient from its own powers
of actuating deeds and words, and whatever is of Love
from its own unselfish and depurated love. The volume
of human language exists ; a man does not take the
volume and analyze it, although he may do so, into its
component points, letters, syllables, and words, but he only
can understand it in virtue of his previously acquired
knowledge, which was, in fact, by a slow analysis and
combining synthesis, and gathering their forms, and the
form of the whole and the underlying movement-forces
which produced it. He intuscepts the volume; he re-

lifes its life-content, or he catches not its life. He has taken, in his antecedent processes, the alphabet, the syllables, and the words and points, and ascertained the analytic value of the respective elements, and from this analytic detail he reconstructs the volume. As the self learns, constructs a letter, a word, a sentence intelligibly, it has intelligently for itself made the letter, the word, the sentence, the volume. I. i. 3, 4, 5. But in all volumes more than mere intellectivity has been employed or, is administered to, and some want, desire, gratification in the Self, and for other selves, is embodied. When the Self is learning the A, B, C, it is a dry, hard, intellective process of cognition ; but as the use of knowledge as administering to some want, some desire, some gratification, lower down or higher up, becomes involved in the progress of knowledge and of life, it is lighted up with a new and correlate power, appetizing to the acquisition and the use. Thus, and only thus, it grasps the intelligibilities inwoven and concreted in the symbols by virtue of its own intrinsic intelligible elements. Thus man rises through forms, I. i. 24, to the essential forces which create them ; to the suprasensible, yet always through the Sensible. God reveals himself through his own alphabet, yet man must cowork with God. So man will, particle by particle, piece the rock, stablish the mountain, ideate the streams and the ocean, build the earth, geometrize, in material content and their forces, the star-systems and find their forms, and in his grand and noble synthesis in which he collects the entire analysis, or so much as may be allowed to his intusceptive capacities, into one Central Unity of Objectifying force, directive and controlling Intellectivity, and all-embracing Love, he will give it a personality and a name ; and whatever that name may be, in any language or nomenclature, it will be of some ideation of God, more or less perfect as these processes are complete. And below this it will be materialism, fetichism, or polytheism, with all their want of moral life, or confusion or want of moral responsibilities. But as man cannot wholly comprehend God, because of the limitations set to his own powers, he cannot *redact* Him to formal limitations. I. iv. 30, 31.

21. This procedure, starting from points and passing out.into lines and surfaces, is the perception of sensations by sight and touch, and it may be of taste, hearing, and smelling, and is the rudimentary perception of form, color, sound, savors, and odors. And this is true, although sight is by reflected light, and sounds are vibrations, odors are radiating particles, and taste and touch are of more immediate contact. In touch the process is of the most conscious appreciation, as any one can see who observes the blind seeing over surfaces through this organism ; in sight it becomes consciously appreciable in learning the exact outline, or the component details of any compound ; in hearing, sounds are attributed to different sizes and intensities of the material producing it ; in sight and sound the vibrations of light and of air have become scientifically demonstrated as affecting these particular organisms ; and the effects on the organisms of taste and smell are in like manner, though recondite and more hidden, assignable to quantities and qualities in causes, acting on their organisms, but cognized, not by observing the outward relations of things, as in sight, sound, and touch, but by their quantitative and qualitative effects on these organisms. Quantity and Quality modulate all things. In like manner the sensations arising from the internal senses, hunger, thirst, erotic desire, motions of nature, are from modifications of their specific and peculiar organisms, depending on the volume and intensification of

the agencies affecting them and the conditions of their
appropriate sense-bearers — their afferent nerves carrying
their special demands in to the common point of perception ; so it must be seen that the spontaneities of man in
his soul-organisms radiate their special functionalized
powers in to this Self, which receives the respective
telegrams of the whole of these moving forces, and from
this centre, through its efferent nerves, it issues its considerate decrees of respondent action or of restraint of
action in breaking up the connection between the animalistic and human impulsions and those other organisms
of its system by which these impulsions are ·executed,
or by which they are retained in the mind for guilty or
innocent contemplation.   In the impulsive action and
conduct of human life, in which this circle of respondent
action is performed without any conscious and determinate
break in the movement, there is the correspondent equivalent of the internuncial circle of the lower orders.   The
philosophy and science of these self-analytic perceptions
and of these considerate decrees, in their notations outwardly, and in controlling and modifying these repercussions inwardly, belong to the consideration of a conscious
and consciously self-determining Self, and this of its Intellective power, yet as of determining between this and
that gratification, and by this or that means of actuation,
and in its time and place of action. I. i..30–34.

22. Intusception will, now and again, be understood
as meaning and as being that process by which the Self,
made self-conscious in the tribal education of life, and in
its own reflex self-culture, goes, as it were, from its own
centre of perception — of cognition — into nature or into
the impulsions of its own organisms, and catching them
in these outward sensations, in the animalistic impulsions,
its psychical psytations, and its own conscious notations

of direction, restraint, or control. It thus catches the *momenta* moving other selves, and in the discipline, instruction, and education of life normalates its own life and its system of the universe — whether as the brute it dies as an animal, or as a mere human instrumentality of its passions and undepurated love, or aspires to its likeness with God and its fellowship with the pure in heart, who see and shall see the kingdom of God. It thus beholds the Forces of life, not as in a mirror, nor as in the dimly illumined cave of Plato, but as in this repeated, self-produced, self-*re*-created phenomena of nature and life, and wherein it sees itself on a straight and narrow way ascending upwards, or with retrogressive steps lost in the devious by-paths of its many animalistic and human appetencies and desires, leading astray.

23. In thus intuscepting the fundamental and initiative momenta of itself, the Self cognizes the exercise of three classes of powers which eventualize as forces, and which are still further analyzable into three simple discrete elementary powers. It will be shown more fully hereafter, and it is now affirmed in the conclusions already attained, that all forces, dynamic, plastic, autonomic, and autopsic, as well as the mechanical, physical forces used by the Self, or in combined machinery, result from the three metaphysical forces herein brought forth to light, — namely, in going out from the centre or Self, or drawing to this centre, or guiding, shaping these two forces around the centre. To limit the statement and yet suggest the illustration, for the present, it will be seen, by bringing the Self sharply to analyze its own processes and its operations, that the mechanical powers are the effects of forces projected from and retracted to a centre, and by an intercorrelating force controlled into a curve or the circle. The only mechanical forces are the projectile, centrifugal,

— the retractile, centripetal, — and the curve or orbital. The projectile force, in and of itself, is a force projecting in a straight line forever ; the retractile, attractive force, draws in a straight line to the centre. The equilibrium of these forces is a stationary, stabilitated point. The libration of these forces, under a directive force controlling both and uniting with the one and the other, produces the circle. I. i. 8–10, 17–29 ; ii. 6, 16, 17 ; iii. 1, 3, 20–25 ; iv. 3, 4, 13, 27 ; II. viii. The attention of the Self, in the matter under consideration, is seen in the mind, being projected outwardly towards nature, and receiving its telegrams from its various orgasmic impulsions, and going into them in their most remote location in the system, and there and from thence cognizing them in their effects on the viscera and in notating its considerate decrees to its animalistic and human impulsions, and. outwardly in nature and in life. It is spontaneous in anger, self-defence, &c. The holding together, the retaining, the centralizing in the self, or of points or masses or of parts, is a retractile, unifying, bringing together, a centralizing force, and in the conscious Self is an attracting, affective force for use, construction, preservation, or dominion. The directive, shaping, redactive force becomes consciously manifest in its control over the other two named forces when using them as physical, mechanical forces, I. iv. 13, and in the determinate control of spontaneous anger, or fashioning it in determinate means of self-defence, and in the control of the gratifications — the attractive appetencies. These three forces in combination make the perfect circle, and in the higher balancing and movement of these forces in the Self, in the moral life, they make the perfect man.

24. Analysis is diremptive, and is the taking asunder, either in actual or mental processes. It is a separating into parts, a projecting and positing in detail for use or

contemplation; it is the unfolding, diffusing of the Self.
It is objective — objectifying. It is represented in personal judgments by all objectivities, as out, outwardly,
thou, thine, theirs, his, hers, its; it is objective, discursive,
diversity, profligacy, squandering, — centrifugal, throwing
or putting off from the self-centre, from any and every
centre. Synthesis is drawing together, collecting, en-
folding, holding together, centralizing, retaining in memory; it is centripetal. And it is represented in personal
judgments by in, inwardly, I, mine, conscious identity,
secrecy, selfishness, centrality. Man gathers knowledge,
and as he gathers knowledge there is in most natures of
this kind, infolding higher forms and more auspicious
organisms, a tendency to diffuse it; in the former collecting, there is a centrality drawing to the Self, and in
the latter a projectility tending to diffusion. The affective — affectionate man loves and *embraces* his wife; he
gathers his children on his knees and around his hearth;
he adds acre to acre that he may be alone in the land, or
that he may subject to his love of dominion; the miser
centripetates; the prodigal centrifugates; in the good
old English, he is a " scatterling." In all men, the
equanimity of character, wherein the well-balancing of
the actuating and the affective forces is observable, will
be attributable to and be found to consist in an equipol-
lency of these forces, which, under the determinating
guidance of the conscious Self by the controlling power
of the intellective force, equipoises and adjusts the force
which projects, diffuses, disintegrates, disunites, loosens,
and destroys with that force which restrains, retracts,
centripetates, holds, maintains, memorizes. It correlates
both forces, and all unite in normal action. Analysis is,
fundamentally, diremptive, taking asunder; in its blind
and uncontrolled action it destroys; it anatomizes man and

the universe when united with the intellectivity in the love and the pursuit of knowledge. Synthesis is collecting, aggregating, bringing and holding together. Redaction is form-designing, means-devising, — the arranging and controlling force. Their joint production supplies forms, organisms, orgasms, and forces to nature and life, and subjected the primordial elements to the dynamic forces, in their times and places, and wheeled the planets around their sun-centres, and the systems of suns in their systems; for simple projection would have sent them out into space in right lines, and mere centripetal force would have drawn to the centre. In all their operations they must move into form, I. i. 24, — symbols, creeds, philosophies, governments, institutions, planets, suns, and systems. I. i. 29. In man they are the Microcosm, and they are the movement-forces of the Macrocosm. In their creative manifestation they are the one birth of time, and were born — inwoven into the first created thing; nay, they are the eternal coördinations, and were the prime potentialities of the Almighty. III. iv.

25. Self-analysis is the only means and furnishes the only method for cognizing phenomena, as it begins with the discrimination of that which is the Self and that which is the Not-Self, and hence furnishes the grounds for the affirmation of the Subjective Identities and the Objective Identities. I. iii. 17, 18. By modifications of its own Self, and by its own modification of matter and forces, it is enabled to affirm its own powers, and the quantities and qualities of matter and forces. And by the modifications of its own continuous life, from itself and from the forces of nature and life, and its own modifications of the lives of individuals and humanity around it, it gathers the fact of a modifiable Humanity arising in its solidaric life, by disenvelopment out of the geotic

and tribal conditions of its early history or lower condition. To return to the phenomena of nature. If the forces of these phenomena and life are not discovered from within and eliminated by its processes, then to it effects are but sequences of unknown causations, and no law or forces as above matter, or as giving intelligible sequences from intelligible causes, can be discovered exteriorily outside of the Central Self which receives only sensations from elsewhere. Whatever else is given must be necessarily received from its own supersensible side. It is seen: *a.* That sensations from the material world are registrated inwardly, through organisms of similar construction, common alike to man and animals. *b.* That ascending higher in perfection of organisms there are animal appetites and desires inwoven in the human organization ; and that these have the same tendency to repercussion as in the flytrap and the polype, and also as in the higher animal organization, having a more highly endowed brain-organism. *c.* That still ascending there are psychical phenomena peculiar to man, yet of kindred character, in many respects, with those of the more highly endowed animal functions, and with higher correlations to nature and life. *d.* That still higher there is a point or centre for the origination or disposal of forces from within to control all the subsidiary, and, in a moral life, subordinated forces at play and work within this complex organization, and to project outwardly, in moral action, into the currents of life, and make the sum of history and of individual responsibilities. *e.* That the Insistent Truth, as it has been rigidly defined, I. i. 35, is given directly by Intuition, as of the native, elementary constitution or essential nature of the Intellectivity. I. i. 20, 23, 24, 40; iii. 9, 29; iv. 7–14, 25. *f.* That the whole complement of forces, in their actual underlying identities and their correlations to one another, and their systematic syntax in nature and life, are only found by the ideative intusceptiveness in its threefold movement from the Self into nature and life. I. iv. 28–31. This may be *perfected* in detail in the complement of all the organisms, yet the law of *the* Life may be attained with some of them — the Blind, Mute, &c. It is not the purpose, nor is it necessary to the conclusions, to analyze the organisms through which the Self manifests itself, and say what material or psychical organism is constructed for the exercise of this or that particular mode of phenomenalization, but as the Self sees by the eye, hears by the ear, gratifies the hunger by the stomach, or other impulsion to gratification by other organs, walks by the legs, strikes or lifts or works or shapes by the hand, or moulds the statue or actualizes the poem in imperishable letters with fingers made cunning by the intellectivity and the love of order and beauty ; or how in hemiplegia the affective emotions are demonstrated, when the will, the actuous power, cannot act determinately and demonstrate itself, I. iii. 6, 7, 9, 11 ; nor by what organisms and intercorrelations of the whole the Self, to so great an extent, obtains the mastery over all the animalistic and human powers, and normalates them into the living and sparkling and more or less transparent symbol of the Spiritual Life, yet it has a whole of organized life.

26. The Self, then, self-analyzing, intuscepting, goes into its sensated or *effected* organism and reproduces a sensation or effect that has been registered, and which can be more or less clearly reproduced, thus calling it to memory, and thus forms or reproduces what may be, and by some authors (Mansel) is called, an Imaginate : —

> "Music, when sweet voices die,
> *Vibrates* in the memory.
> Odors, when sweet violets sicken,
> *Live* within the Sense they *quicken.*"

In every Thought there are three elements, namely, — *a*, The Thinking Self. *b*. The object about which the thinking Self is engaged. *c*. The Imaginate, the figure, the form and quality, which has affected its appropriate organism and left its modification in the memory. The latter is the link of mediation which unites the two former, absent or present. That this is so when absent, will be readily perceived. That this is so when the object is present becomes evident when it is seen that in successive periods of time, in every identification of the object in the memory, in every successive moment, a memory of the object in past time is necessary for its identification in the present; and this is but a comparison or continuation of the original impression to such present time. Brewster, in his "Letters on Natural Magic," shows that when a natural object of sight is recalled to the memory, it is *repainted* on the retina of the eye. Interior observation of the internal senses, and of the passions and affections, will show, satisfactorily, that this is the case with each of them, as thinking of each tends to produce each, as in appetites, anger, love, &c. And this tends to explain all those sympathetic movements of which man is the subject, and in so many ways the victim, in the preponderance of his animalistic and human orgasms, wherein the presence of their correlated objects in nature and the antagonisms of life excite the respondent passions and desires for gratification. § 10. Such are the penalties of improper early impressions; and in the bland and genial impressment of the Forms of culture, truth, and holy life on the fresh minds, these terrible orgasmic forces may be modified, controlled, and, by the unfolding love of the higher life, subjugated. Imaginates, good or sinful or morally indifferent, are thus the reproductions, in greater or less exactness or force, of the impressions previously or habitually made. In their determinate reproduction it is shown that the Self, from its own interior centre, projects, intrudes itself into the organism affected or the respondent viscera to which it has been repercussed — reflexed, and reproduces the sensational images or impulsions. When this is not determinately done, it is produced by the sympathy of the passions or affections and their repercussive interactions, or by the spontaneous action of the registrated organism itself when the overdue tension of the intentive Self in trying to recall is withdrawn. §§ 9–11.

When the Self contemplates an object, the Imaginate of that object appears in the appropriate and correspondent organism. These various organisms are frequently seen to act automatically or spontaneously, as in reveries and impulsions of the desires and passions, and in these last two instances like most of the cases of the unnormalated life when the violence of the animalistic or human orgasms is the rule rather than the exception of conduct, as is nearly always the action of the two spontaneous forces, I. i. 17–23 ; yet they are susceptible, to a greater or less extent, to culture or suppression, and are then brought into direct and conscious regulative correlation with the autopsic Self. Yet in the most of life it will be seen, when closely inspected, how little of its action is from the direct and independent Self in its disenveloped solidaric independency, ruling these orgasmic powers, strengthened as they are by constant exercise in the indulgences and pursuits of life. The proper independent Self, in the sanctitude of its spiritual, solidaric freedom, is almost relegated from the domain of active life, or is made the instrument and the slave of the animalistic and the human passions and affections. I. i. 32.   As the conditions of the human selves approach

15

that point where the Proper Self is incapable of exercising any clear independency of action, it may be expected
that then the organisms will be more likely to act automatically and upon foreign influences, as in hypnotized,
mesmerized subjects and *rapporting* mediums. The
independence and the dependence of the Self in its
psychical organism may be illustrated from facts sufficiently established to find a place in a standard work
on Physiology. With the explanation, that, in the
delicacy of organism which is necessary to give the Self
the power of acting on the various organisms for its own
manifestation and their control, these organisms may
become so automacized or so inordinately excited and,
in the instances referred to by Carpenter, so hypnotized,
that they may act independently, and in extreme cases,
as in manias, &c., react upon the Self and prevent its
normal, autopsic manifestation and control. I. iii. 15.
And it will prove in these cases, as in all cases of continued indulgence, of a yielding up by the proper Self
to the demands of the animalistic and the human appetites,
passions, and affections for their natural orgasmic gratifications, that they will more constantly prevail, until
this proper Self sinks and disappears from its regulative
supremacy. In the instances hereinafter given, it will
be seen that the hypnotic effects produced were in those
organisms which had been frequently used, and that
they were only the old currents in their registrated lines
re-excited into independent, and as it were in their local
organic action, and they will serve to show the *biologic*
action which involves communities in the madness of
popular infuriations, and sets in high moral relief the
nobler and better portions of society who are able to *conserve* their powers and maintain their moral independence
against bloody and remorseless or wildly extravagant

fanaticisms. They show that these temporary hypno-
tisms or mesmerisms are in the nature of temporary insanities, or that monomania is like a mesmeric state impressing a particular function with a line of thought or
a state of feeling. Carpenter says, " Artificial somnambulism induced by the (so - called) Mesmeric process,
(§ 696,) or by the fixed gaze *at a near object,* (as practised
by Mr. Braid under the name of *Hypnotism,*) is essentially the same as that of the ' biological ' condition, save
in the different relations which they respectively bear
to the waking state ; for there is the same readiness to
receive new impressions through the senses, (the visual
sense, however, being generally in *abeyance,*) and the
same want of persistence in any one train of ideas, the
direction of the thoughts being actively determined by
the suggestions which are *introduced from without.* In
either of these extreme forms of Somnambulism, and in
the numerous intermediate places which connect the two,
the consciousness (?) seems entirely given up to the one
impression which is operating upon it at the time ; so
that, whilst the attention is exclusively directed upon
any one object, whether actually perceived through the
senses or brought suggestively before the mind by previous ideas, nothing else is felt. Thus there may be
complete insensibility to bodily pain, the somnambulist's
whole attention being given to what is passing in his
mind ; yet, in an instant, by directing his attention to the
organs of sense, the anæsthesia " (this isolated condition
of the functional action) " may be replaced by ordinary
sensibility ; or, by the fixation of the attention on any one
class of sensations, these shall be perceived with most
extraordinary acuteness, while there may be a state
of complete insensibility as regards the rest. Thus
the Author has witnessed a case in which such an exal-

tation of the sense of Smell was manifested that the
subject of it discovered without difficulty the owner of a
glove placed in his hand in an assembly of fifty or sixty
persons ; and in the same case, as in many others, there
was a similar exaltation of the sense of Temperature.
The exaltation of the Muscular Sense, by which various
actions that require the guidance of vision are directed
independently of it, is a phenomenon common to the
' mesmeric ' with various other forms of artificial as well
as of natural somnambulism.   The author has repeatedly
seen Mr. Braid's ' hypnotized ' subjects write with the
most perfect regularity, when an opaque screen was interposed between their eyes and the paper, the lines
being equidistant and parallel ; and it is not' uncommon
for the writer to carry back his pen or pencil to dot an i
or cross a t, or make some other correction in a letter or
word.   Mr. B. had one patient who would thus go back
and correct with accuracy the writing on a whole page
of note-paper ; but if the paper was moved from thé
position it had previously occupied on the table, all the
corrections were on the *wrong* points of the paper as
regarded the *actual* place of the writing, though on the
right points as regarded its previous place ; sometimes,
however, he would take a fresh departure, by feeling for
the upper left-hand corner of the paper, and all his corrections were then made in the right positions, notwith-
standing the displacement of the paper.   So, again,
when the attention of the somnambulist is fixed upon a
certain train of thought, whatever may be spoken in
harmony with this is heard and appreciated ; but what
has no relation to it or is in discordance with it, is entirely
disregarded.

   " It is among the most curious of the numerous facts
which Mr. Braid's investigations upon artificial Somnam-

bulism have brought to light, that the suggestions from
the Muscular Sense have a peculiar potency in determining the current of thought. For if the face, body, or
limbs be brought into the attitude that is expressive of
any particular emotion, or corresponds with that in which
it would be placed for the performance of any voluntary
action, the corresponding mental state, that is either
an Emotional condition affecting the general direction
of the thoughts or the idea of a particular action, is
called up in respondence to it. Thus, if the hand be
placed upon the vertex, the Somnambulist will frequently,
of his own accord, draw his body up to the fullest height
and throw his head slightly back ; his countenance then
assumes an expression of the most lofty pride, and the
whole train of thought is obviously under the domination
of this feeling; as is manifested by the replies which
the individual makes to interrogatories, and by the tone
and manner in which these are delivered. When the
first action *does not of itself call forth the rest*, it is sufficient to straighten the legs and spine, and throw the
head somewhat back, to arouse the emotion, with its corresponding manifestation, in full intensity. If, during
the most complete domination of this emotion, the head
be bent forwards and the limbs be gently flexed, the
most profound humility takes place. So again, if the
angles of the mouth be gently separated from one another, as in laughter, a hilarious disposition is immediately generated ; and this may be made to give place
to moroseness by drawing the eyebrows towards each
other and downwards upon the nose, as in frowning. So
again, if the hands be raised above the head, and the
fingers be fixed upon the palms, the idea of climbing,
swinging, or pulling at a rope *is called up in such as
have been used to such kind of exertion ;* if, on the other

hand, the fingers be flexed when the arm is hanging
down at the side, the idea suggested is that of lifting a
weight: and if the same flexure is made when the arm
is advanced forward in the position of striking a blow,
the idea of fighting is at once aroused, and the Somnambulist is very apt to put it at once into immediate execution." See the same author's notes for other instances
of the same kind, thus verified upon competent evidence
by high scientific authority.

Now remark: a. That in all these instances the controlling self-consciousness, the Conscious Self, is put in
abeyance, and these acts are the automatic manifestations
of independent organisms, which have been charged with
habitudes of action as well as by their intelligential
function of special activity. It is but the excited action
of the special organism in its separate function manifesting the special form of activity with which each is endowed, without the conscious presence or control of the
Self. The acts done under such circumstances are not
done with legal or moral intention, for the determinate
Consciousness is in abeyance. It will explain much of
the spontaneity of human action. I. iii. 11. b. That the
manifestations exhibited are " called up in such as have
been used to such kind of exertion," showing that previous
registration in the organism or original function as inwoven
by nature in the organism is necessary to the reproduction of the manifestation. These phenomena will not
take place with the idiot, nor will an emotion or muscular
action of a definite kind take place in one who has not
previously exhibited the emotion, or been accustomed to
perform the act. It shows how the human organisms
may become automatic; and it tends so explain how
the animal organisms of the brute races may be and are
so constructed with their highly organised and specific

special senses as to act and react on each other by simple
internuncial Repercussion, and without a consciously superintending Spirit, and how in their subjugation to man
they may be more or less hypnotized and made obedient,
more or less permanently, as the animal is organized, to
the controlling energies of man. It shows how easily
strong passions and affections are persuaded or tempted,
I. ii. 15, as it also shows the dangerous tendencies to
religious and fanatical influences among men, themselves
in those states of individual or social life ·where the
organism is unbalanced by controlling or librating organisms, or the culture of life, § 10, and the moral independence of the Self does not exist.

In the delicate machinery of the human organization,
made exceedingly delicate to respond to the direction
and control of the Self, this very delicacy subjects the
various organisms, which are the instrumentalities of the
Self and the means of communicating in to the Self, to
extraneous influences, and they must in turn be subject
to various disturbing causes from without and diseases
and disturbances within. That these are under the
control, to a greater or less degree, by medical treatment,
is also well known, which, in many instances, restores
them to their normal condition of action, and thus the
Self can again act on and through them. To one school
of Medicine these facts are well known, and anguish,
vexation, audacity, obstinacy, despondency, hysteria, &c.,
&c., are the subject of *direct* medical treatment, by operating upon and, in some degree, modifying the condition
of the organism concerned ; while all know that the surest
and only radical remedy is the enthronement of the
Conscious Self in its regard for Rectitude, Order, and
Submission to the Divine System of the government of
God. The importance of the facts, and the distinctions

will appear. I. i. 31, 32, 33; ii. 5, 7, 8, 14, 15; iii. 5–16; Book II. *passim.* They are the reproduced imaginates without the Conscious presence and control of the Self. II. iv. viii.

Pointing out the fact that the psytations as natural impulsions of their respective orgasmic forces only occur in the organisms of the two spontaneous forces, of the passions and the affections, and yet that they are registrated in the viscera as Imaginates, but that the Imaginates of the Intellectivity are not native and instinctive to it, but are gathered through all the avenues of cognition hereinbefore brought into view, and having pointed out how the Imaginates are restored by direct recall and associative sympathies, it may be asked how the unhappy race of mortal men may be improved and elevated? Surely the first step will be to prevent and counteract these vulgar, vicious, and sinful Imaginates which so constantly return and settle into habitudes of the Soul. And this can be done only by giving to the early mind the Forms of Thought animate, glowing, and intelligible with the imperishable Life. Language, as has been intimated, § 24, and as it will be more fully shown to be the germinal growth of the triplicate powers in the Self, is most frequently purely representative of each in its direct and discrete difference, and also of their complex combinations and correlations; and language from the earliest to the present times has shown the general mastery of the animalistic and human impulsions to conduct over the spiritual powers in man, yet throughout they have been separated, in virtue of these antagonisms lying in the very roots of our existence, almost as light and darkness. The animal-man has always had a language and a literature expressive of the powers impelling him to thought and action, and indicating his loves of gratification. It was and is a faithful expression of the animal-man. It expresses directly and implicitly his great lusts, pride, cupidity, and voluptuousness. In form and spirit it is at times fierce, hard, haughty, cold, pragmatical or hypocritical, but more frequently elegant and voluptuous, either to cover the shame of its vulgar and unworthy motives of conduct, or to give attractions to the wretches it would betray, and the poor dupes it would use for its instrumentalities. But the types of the thoughts, the feelings, and the eventual action of the other is essentially spiritual. It is rich, simple, sublime; most elevated, most profound, most chaste. It is solid in its intellectual constructions, and it is brilliant in its glorious adornments. Like the life which produces it, " it is, in turn, rich, spiritual, simple, sublime, true, sweet, chaste, serious, and modest," or glorious in ornament as the richest garniture of the skies, or swells with the very grandeur of the Heavens and the awe and reverence of the majestic Presence; and it is the constant preacher of Spirituality.

> Light and Darkness bless the Lord ;
> Types of the Thoughts which sever —
> Types of good and evil Word —
> His love or justice praise forever.

27. As an Imaginate is thus representative of forms and qualities and movements in nature and life, and of the psychical and spiritual phenomena within, so the Self, in its ascent and educational progress, stands in need of some more general symbols or tokens for races, species, classes, families, — generalizations of individuals of a similar kind or organization in the aggregate. The form of expression must be more inclusive than when speaking of a designate individual or any number of individuals of the same kind or organization. It must include the

whole. Concept is then that mental expression, representing by its arbitrary term the general fact that there are a great many things of the same kind or organization of which we can affirm there is an individual thing, as the man, a man, man, in the last instance, for men generally or for mankind. Some general characteristics of the class are analyzed and grouped in the mind, and there represented, contracted, stenographed in the symbolic term; and each one in the use of the symbol, in so far as he gives it a content, must do so from his own Self, his imaginates or sensates, and in intuitions from his pure intellections, yet as borrowed from actual forms. I. i. 24. From the want of that Speciality and individuality, which is the character of Imaginates, concepts will be the more vague and indistinct.

28. In the investigations of philosophizing it is claimed that there is a something, a somewhat which underlies all phenomenalization, II. iii. ; that phenomena alone are recognized, and can only be made known to the Self as indicated in sections six and seven, while the underlying identities which produce the phenomena remain unknown. The proposition may be exemplified thus : — the tulip, the lily, the hyacinth, and the onion are bulbous roots, similar in appearance, and planted in the same soil, watered by the same moisture, and growing in the same air and light; they yet produce different forms of plants and flowers; but that autonomy in each which with such cunning intelligence and positive exhibition of force takes up the plastic elements of nature, earth, air, water, and light, and produces these different forms of plants and flowers, escapes all the tests of the senses. This base for the autonomy is *notionalized*, as it has been called, and the mental result is called the "notion." The unknown Somewhat, reached in this way, which is the same herein-

before used, I. i. 38–40, 26–29 ; iv. 28, is the differentiate
autonomy. It is, in phenomenal metaphysics, the subsistence — the noumenon for the phenomenology of the
manifestations. Here again it is seen that in every effort
to reach the inner underlying forces it is a process of
going in through the outer covering of forces to their
causations ; and that here again are found the three forces,
in some form of combination — the intelligential force
moulding the particles as they are attracted to and by
the germ and expanding the limits of its form.

29. For the conduct of life, and in reference to the conduct of others, and in considering many of the operations
of nature, e. g. as in the weather, there are numerous
philosophical contingencies, ℟ i. 13, to be taken into consideration in the formation of judgments as to what has
been done or what will be done or come to pass, or who
or what agency did it ; and the process by which these
judgments may be formed are properly called opining —
a bringing together of the philosophical contingencies
involved in the agencies at work, and declaring the
probable issue in the judgment of the thinker, and the
result is *opinion*. It is but the process of ideation applied to the philosophical contingencies of nature and life
in the common occurrences of life.

30. Intuition has already been sufficiently set forth as
the dry and pure logic of the intellectivity disposing of
the correlations of quantities, whether of surfaces or
weights. I. i. 35, 36 ; iii. 29 ; iv. 19, 25. And Ideation
as the triplicate processes for obtaining the Divine Ideas,
and the Proleptic Morality, the Transcendental Being,
and his prolepsis for the movement and order of the
Cosmos. I. i. 36, 37–43 ; ii. 2, 3 ; iii. 29 ; iv. 3, 4, 18,
19, 20, 28 ; *ante*, § 16.

31. The processes brought into view seem mainly to

refer to the exercise and manifestation of the Intellectivity; but while it is theoretically separable in pure
Formal Logic from its cognate correlative powers, yet it
is in actual life always connected with them, for it is
exercised for some use which involves them. But at
every turn of the processes the two other distinct elements,
subsistences as forces are brought into view, and cannot,
without violence to the whole system of the universe, be
disregarded, based as they are on essential differences in
their phenomenalizations in physical nature, and performing their offices in the moral economies of the individual, and in human and in animal life. They are, as
so often has been brought into discriminate notice, the
Actuous, the Objectifying, Objectiv-facient Force, and
. the Attractive, the affective force of all temptations,
solicitations for the things of nature correlated to its
action, and thus the Self to them; and this Love is the
inherent Appetency of all life, the Attractive force of all
nature. And these two latter elements are not only
found in the actual, empirical psychology, but not a step
can be taken in a completed Rationalistic Psychology
without their prompt and efficient recognition. When
once possessed by the Self, they are like its own Consciousness, they cannot be denied; and every effort to
remove them from the realms of nature and life will
make them irradiate more and more their light and their
place in the universal system. I. iv.

32. The Actuous Force is that objectifying power without which Nature cannot be conceived as having been
set over in objectivity from its creator, or as having any
immanent existence. It is the power by which the Self
is carried, and in conjunction with the intellectivity is
determinately sent out into overt action; and it manifests
itself in spontaneous and thus in indeterminate movements.

It demonstrates itself with greater or less power, from the expression in the eye or on the brow or lip, or in the terrible blow of the convulsed arm and hand ; and is modulated by the intellectual culture and love of art in moulding the statue, the poem, the book, and all works of art, and in the civilizations and in the evangelization of life. Yet in all it is seen in its correlations with its cognate forces, and as deriving intensification and direction from them, except in its simple spontaneities of anger and self-defence, &c., and it loses much of this characteristic in the culture of life, and still more in the spiritual subordination which binds it with the chains of a holy Love.

33. Love is a term, perhaps, of as much indefiniteness, yet always definite, as any one in the language. Between the sexes it is frequently the synonym of lust; in the citizen, it is patriotism ; in the parents and children, it is affection ; in marriage, it is the foundational element on which all the institutions of Christianity are founded, and without which Christianity itself is a rope of sand, having no ligature to bind families into a holy solidarity, parents to children, children to parents, and all to society, and society together. Yet in the mob and animalistic populace, or mere human associations, it is an indefinite impulse which produces the hurrah for this leader or that, this party or that, this flag or that, and as such is a delusion and a folly, it may be, crime, when it observes secular forms which are comparatively indifferent, and yet forgets the weightier matters of the moral coherencies which should bind man in families, governments, society, and the evangelism of the universal society, based on the solidarity of the races, yet in the moral correlations of their tribal and historical conditions. In the gourmand it is gluttony, in the miser it is avarice, in mere human government it

is centrality and despotism, in the evangelist it is charity, and in Christianity it is the universal brotherhood of the races, yet under the conditionings of the Prolepsis of the Almighty. Throughout the entire range of natural appetencies and moral attractions it is Love, until we reach the comprehensive and all-comprehending Charity, and thence truly attain the great height where we learn in the conscious and unfolded yearning of the Spiritual Self, that " God so loved the world" that in the consummation of his ages " the fruit of the Spirit is Love," and thence feel inclined, nay, are attracted and drawn to look profoundly into nature and life and see what hidden and yet open element in the constitution of things it is which binds all things together, and makes physical causes and effects the instrumentalities of the moral life, and binds in the normalation of all languages all these seemingly heterogeneous desires for gratification and the affinities in nature under one common appropriate designation of attraction, and which under all forms of expression, whether of appetency, sense of gratification in a sense to be gratified, temptation, solicitation, or love, is only understood and appreciated fully when it is seen and felt as an attractive force, showing ever in this its intelligible language, its action, and its correlations in nature and life. In its highest ideation, this movement in the spiritual manifestation of Deity is best represented by the Greek word, $o\rho\mu\eta$ — orma, translated into Latin by the terms *impetus, impetus animi,* movement of the mind, and derived from $o\rho\alpha\omega$, *orao,* to see, to understand, a discerning love. And so the Ormaic Love is occasionally used, when speaking technically, of the coördinating force, the third coessential element of Deity. I. i. 8–10, 17; iv. 3–16, 20, 21, 28–31. It is necessary to complete the ideation of any act of creation by the Deity, whether it is viewed as the exercise of physical power or as the movement of moral Personality. In the Self it is found in the Consciousness, and no normal moral act is conceivable without it. I. i. 17 ; ii. 13 ; iii. 9. This third coördinating power is the ormaic essence in Deity in virtue of which Love is the base of gratification in nature and life, and the *attracting* end of all spiritual aspiration, both in its own intrinsic nature and the moral necessity for reciprocation. Love must have Love returned. They correlate each other, and the necessity of thought requires both. I. i. 15. So Love is the base of the affectional emotions in man, and is in him personative and representative of the divine creative *orma*. This love in man, in its unfallen or its disenveloped and depurated normalation, is seen as the profound unselfish and purified love, such as a Spirit must have for a spirit where there are no fleshly, earthly environments of attraction in its many forms of interests, wants, desires, appetites, comforts, conveniences, and enjoyments of the animalistic and the human organizations. Such love, in its own entire self-abnegation from these clinging impurities, such as the sainted living have for the sainted dead, would be, in the complete Self, an impulse and a principle of goodness, and thus not only the image, but the likeness of God. Who will ascend, who will *intuscept* with Christ, the great Teacher of Life, in this the only Method of attaining it, into this region of holy, contemplative, and active Love ? II. viii.

34. This gives three entities — three diversities in the Self. Into these three diversities the Self can enter and thus intuscept them. It can stand by the Actuous power and interrogate the Intellectivity, and demand from its cognition of the whole realm of organized existence the exhibition of its whole fund of cognitions and fancy and argument for what motive-love it should actuate ; it can

stand with the Intellectivity, front to front, to this same motive actuator, and demand that for such and such motive end of conduct, and by such means and in such manner, it shall be projected into actuation, and execute and enjoy or vindicate motive, and in vindicating it, enjoy it; while it turns to the loves and selects its worthy or unworthy motive-cause, and finds its cause to actuation in this or that causative-end of gratification. I. i. 15. With the Love it can stand towards its lowest phases of appetites, desires, wants, and passions, and too often successfully demand the beastly or the human gratification, or, in a higher ascent, seek artistic enjoyment and intellectual gratifications, and, aspiring to a still higher excellence in the fuller normalation of its life from its intuitions and ideations, catch the moral effluence raying out through the realms of nature and life, and, in the conjugate exercise of its conjugate Powers, it can, in the method of the great Teacher, intuscept and feel and in some sort live the Beatifications of the Pure in Heart. This gives the unfoldings and the enjoyments of the progressive, the intusceptive, the threefold life-moving Self. In the line of its infinite progress, it is the conscious self-accumulation and exhibition of those powers which, in themselves, contain the *means and the end of the progress*, yet only as they are accorded to the unfolding love made worthy of the Love which accords. This leads life to sovereign Moral Power, and the action and endurance of these powers sinew the Soul for higher action, and the Triplicate Will, circling through all experiences, unfolds into the cognition and actualising appreciation that Pure Moral Law is perfect Freedom.

35. It is becoming apparent, that, starting from the interior of the Self — from the self-centre, in the perception of sensations, in the cognition of their differences,

in the analyzation of special details, in synthesizing the
details into articulated wholes, and joining such wholes
into a syntactic system by their filamentary correlations
in the formation of opinions, notions, intuitions, and
ideations, attaining the Divine Ideas and the Proleptic
Morality, indicating a Creator coördinated in an Immutable Morality, to which all these supertend, there is
an *egressus*, a going forth of the Self, and there is a
gradus, an ascent as stated by the philosophic Hermes.
§ 1. This *gradus* is through the beholding by the Self,
through sensations, imaginates, thoughts, judgments, concepts, opinions, notions, intuitions and ideations ; and in
conjunction with these there is a gradation of life — of
living movements — through instincts, passions, and affections, to the purified love, carrying this Self into its actuation into life by a new knowledge gained in the ascent
of the gradations through animal and man up to the angel
standing by the Throne of Everlasting Love, and revealing to the ascending Self, in his exaltation, the likeness
of the triplicate image to the coördinated Trinity in its
Unity of law and order ruling into life, the life of the
infinitesimal protoza, and unfolding the love in conscious
races which is incomplete and abnormal without a reciprocating Love. The Self, making from its own centre its
egressus, by the education of life and the Grace of God
— his Love reciprocating — passes over each step of the
ascending *gradus*.

36. A main purpose therefore has been and will be,
yet further, to establish the separate, distinct identity of
the threefold correlate elements in the Self, their mutuality and their dependent correlations on each other
and to nature and to God, and that they are personative
and representative of trinal essences in Deity, which
subsist each to the other, each with the other, in neces-
16.

sarily eternal coördinate coessentialities. As we can see
the three correlate elements in man exhumed from this
envelopment in the environment of their animalistic and
human orgasms, and their proleptic and progressive
exaltation in the restored condition of their depurated
activities, the Self will become sinewed with action for
further action, in the redintegration of the living forces
which in the fulness of life will give the likeness of Him
which created and who is the fountain of all life.

37. This is reached by the *egressus* and *gradus*, and is,
throughout the ·ascending gradation, the intusception of
the Self and through the Self of nature and to the triplicate ideation of the All-Mighty, All-Wise, All-Loving
God. It is, throughout, in every step of actual progress,
an analysis, a concurrent synthesis, and a final redaction
into completed form, yet, throughout, by subsidiary and
ascending forms. I. i. 29, 23, 24, 38–42. The diremptive analysis breaks up and disintegrates the solid wall
of life and existences which separate the Self from the
suprasensibly Actual — the spiritual — the Creative
Forces. And then the synthesis, from the broken and
disintegrated materials of disorganized forms, dissolved
organisms, powers spent in action yet never lost and continually renewed in the prescribed economies of the correlations of life and nature, functional forms and orgasms
exhausting and exhausted, yet their *débris* furnishing
sustenance to new vitalities and onward movements,
gathers the laws and the forces of normalation by which
to *re*-form, redact the arch which must span the gulf
that separates man from beholding the spiritual forces in
their intrinsic identities. And this arch is the pathway
which leads the progressive soul over the abyss of nature
to God.

38. Empirical Psychology is based on the facts of

Consciousness rigidly ascertained and accurately defined, and each referred to its appropriate place and connection — correlation in the Self. And Rational Psychology claims to " lie originally in a very different field from experience," and " to seek for the rationale of experience itself in the necessary and universal which must be conditioned for all the facts of a possible experience." — Hick. *Rat. Psy.*, 1, 2. This is not true nor scientific, unless received with limitations which will confine it to the Formal Logic, or necessitate the inclusion and consideration of elemental subsistences which cannot be found in mere intellection, and which must come from experience in the movement-phenomena of other intrinsic forces. These limitations are : *a.* The only necessary and universal which is contained in a mere Rational Philosophy — any Rationalism, is and only can be the Insistent Truth. I. i. 35. This has been and will be further shown to be only of forms of quantity, and as such are necessary and universal. *b.* The Immutable Morality of the Divine Self, while it must be postulated and affirmed, cannot be circumscribed in any forms of human thought, I. i. 36, but the Divine Ideas must be seen as adjustable forms for creation and adaptable and adapted to the selected place for this planet in its place among the planets, and therefore as neither necessary nor universal. I. i. 36, 41, 42, 29, 25 ; ii. 19 ; iii. 1, 28-31 ; iv. *c.* In like manner it has been and will be more completely and conclusively shown that the Proleptic Morality is a special appointed or statutory law for the government and progress of men in their specific autonomies on this planetary theatre of action, I. i. 35, iii. 1 ; and also as not necessary and universal, except with reference to creatures so organized as man in a theatre of life and action such as this world, and moving onward to a theatre of higher or other action.　d. That life and nature, as empirically known — experimentally seen, can only be transcendentalized as from a creative source and by a representative Self having Actuous power, Intellectivity, and Love, and from its point of transcendentalization feels and lives and knows these forces as they are inwoven in the movements of creation. The most favorable judgment which can be allowed to Rational Psychology, I. iii. 31, is, that it is but a restatement of the Empirical Psychology from the synthetic side, after the empirical analysis had ascertained and defined the facts, and referred each fact, by the concurrent synthesis, to its appropriate place and connections. In this sense, in rational psychology, the facts which seem to be abnormal, eccentric, and individual, as well as those which are common to the race, are to be rigidly ascertained and restored to their systematic unity in the relations of time and place and exact history of their appearance, together with their more or less gradual normalation into the life of a people and their formulation into language; and thus, under the processes instituted, the concrete, the actually conditioned, may be seen in itself and its correlations, lying, as it were, in the precise matrix of the creative formations; and this will give spontaneity and a progressive normalation of life. As these types and ectypes, and their constituting and constitutive forces, are pursued by the Self, it will intuscept nature and life otherwise than by an intellectual anatomy, which only kills the living life to find and get the moveless skeleton. It is manifest that there is a large class of minds which cannot see philosophically; cannot grasp the ante-typal idea in itself, nor, consequently, in its philosophic and appropriately appointed correlations from any number of facts, however great; I. iv.; ante, § 18; and that another can only grasp these

general laws, executing themselves in law-forces, from
an accumulation of direct facts and the comparison of
collateral facts ; so there are synthetic minds, which,
from a comparatively few facts, will boldly reach at the
general law, the transcendental ideas and the forces for
executing them from the Initiate Causations and as
they are inwrought and inwoven into the *facta* — the
things made. Empirical philosophy, when consciously
conducted by the triplicate activities of the Self, gives
the facts, and shows how the facts actually correlate and
adjust one with another; and as they are seen and felt to
be intelligible in this very intelligibility, they require the
intellectual foreplan which ruled the phenomenalization
of the facts. A full psychology gives the intellectual
foreplan, and in the intusceptive ideation of the divine
forces, functionalized into the law-forces at work, it gives
the divine prolepsis and traces the facts as they appear,
flow from — are determinately produced and actualized
from the primordial causations. The rational psychology,
if as such it is at all possible, is purely formal, and as
lifeless as the pure figures of geometry ; the empirical
psychology, when conducted as a pure intellection, is but
a learning of nature and life from its under side towards
man — an *a posteriori* knowledge, I. iv. 3, 4; while the
a priori cognition of the Divine Ideas, the Proleptic
Morality, and of the movement-forces of nature and life,
and the self-possession of the whole which makes these
intelligible, makes him who reaches this union of the
divine and human, a son of God, and crowns him with
Light and Love. But in this long travel there are
Teachers and Learners, docile and indocile, the obedient
and those who *must* be submissive.

39. The only perfectly synthetic mind is Deity. In
his transcendental Intelligence, in his Omniscient In-

tellectivity, lay the universal cosmos in its divine, unact-
ualized, unmaterialized foreplan — in its unsymbolized
idea. This was fashioned forth in the multitudinous,
complex, yet consistent and correlate details of creation.
In' this proleptic idea was, is, the absolute synthesis.
This synthetic state, in the actual movement of the
coördinate forces, descended into the practical, into the
material, the concrete. God descends from synthesis —
his omniscience — into his analysis, the creation of things
in detail; man ascends, can only ascend from analysis to
synthesis, in whatever formal manner he may declare
the processes and the results. Creation exists only in
details; it can only so exist, and man can only catch the
synthesis, part by part, in his fragmentary analysis of
the details. This divinely concrete, in its actual combinations and adjustable correlations, is presented to man
under various synthetic conditions, as water, air, bodies
and instincts of animals and souls of men, and in less or
greater systems; but their preëstablished correlations —
preëstablished in the very conception of their adjustabilities — bind the whole into a determinately arranged system. Man must, he only can, by slow analysis solve this
natural, and ascend to the spiritual synthesis, that is,
analytically reconstruct, re-form the synthesis. These
he does only in virtue of the Intelligibilities inwrought
in the concrete. These Intelligibilities are intelligible
to him only in so far as he sees that which he himself
possesses; he understands actuation in others or in
Deity only in virtue of his own power of actuating. So
of the intelligential workship in plants and instincts in
animals, and so of the intelligence of men or angels or
God. So of the affections in their diversities of functionalizations, and in the simplicity and singleness of
the Love which is the base of all gratifications. § 35.

The physical law-forces are on the outside toward and in nature, and they are the resultants of moral forces; hence physical life must be first comprehended or apprehended before the moral life can be reached. We must go back through the acts and conduct of man to the moral life of the man, and through nature to God.

40. Starting from the ground of sensation in the organisms, and passing through the *gradus* of cognitions in perception and their colligations in opining, notionalizing, intuitating, and ideating, and embracing the conscious actuations and affections, and claiming that, now in the history of the human consciousness, and as it can now be observed in the gradations of tribes and nations, there is a historical unfolding of the solidaric Self, and that, as it unfolds in the complement of all its powers, in its deobscuration from its organic autonomy, it comes to know and love God, and by the very sympathies of these powers, in actuating its life, seeks order, justice, righteousness, and eschews Fraud and Force, the twin children of Iniquity. Finding that the synthesis of, nature and life in their mutuality and difference, and in the preparation and adjustment of physical facts and forces to moral agents and moral forces, has been given to man for the exercise of his highest powers, and a method has been bestowed upon him by the great Teacher wherein he can attain the highest law of self-government for the moulding of his own actual positive life, and from him-self, in some degree, impart new life to others, the foundation is laid for binding the whole into one syntactic system in the reconstruction of the universe from the Power, Wisdom, and Love of God as living, coördinate Forces, and radiating their triple light in our own lives and into the great current of human life.

41. The harmony of all views which are true is

their just and syntactic combination by a view or method common to all truth; and Morals and Physics, Mind and Matter, must be seen to be harmonious by *a posteriori laws* or by *a priori forces* ruling and ordinating both in their very creations, and arranging the correlations for their adjustabilities, or in some theory of Emanation, Transformation, or Development. Nature is given to man in greater or less synthetic detail, and the Thoughts of a Holy Wisdom have been gathering in the ages in words of light; and from the very nature of the facts and circumstances, the times, the places, the agencies employed, some of the facts exhibited and doctrines enounced with greater simplicity or clearness at different times and by different persons, it is the province and the duty of the Self, standing on this parallelism of facts joined together by the very laws and mutuality of forces which unite or intercorrelate the moral and the physical natures, to reach forth, through the broken, disintegrated and ever-changing materialism of nature and the confusions of creeds and inconsistencies of faith and practice, and the superstitions common and native to the human mind, and the constant tendencies of mankind to malversate the highest and holiest forms of Truth to the gratifications of their human passions and affections, and thus return, more or less degradingly, to their native conditions, and, passing through all these, reach to an ultimate synthesis, and, consequently, to a concordant philosophy and a coördinate Religion. In each step — and the appeal is to the intelligent consciousness — new and brighter and serener light will irradiate upon the path from the intelligibilities within and beyond, and which are inwoven and implexed in nature and life; and each new ray of light, as it is gathered, will increase the volume of light, until, in the highest upward progress, all the stones and beams and apartments of the great superstructure of the Temple of the Universe will be seen to be bathed in the effulgence of Light and Love, and the Stars of the Morning, which sang for joy in glory of their creation, in their Living Forces, still hymn their Praise.

BOOK FIRST.

BUILDING-STONES.

—◆—

CHAPTER SIXTH.

MATERIALS OF CONSTRUCTION IN FURTHER DEFI-
NITIONS.

1. ALL philosophical investigation and all positive
knowledge in formulas of science, as well as in popular
and common acquirement of information, belong to one
or other of two chief divisions — Physics or Pneumatics
— Πνευμα, Spirit, in its threefold spiritual meaning.

2. Physical Causes are the result or the determinate
effects — production of Moral Causes, I. i. 8–10, 15, 16,
18, 19–25, 41, 42 ; ii. 2, 19 ; iii. 1, 19, 20, 25, 29–31.;
iv ; v. 15–17 ; and as physical causes are seen to correlate and specially *effect* different organs or orgasmic
powers, tempting human passions and appetites, and
soliciting the various distinctive instincts of the animal
and human natures, I. i. 30 ; ii. 11, 12, 13, 16, 18 ; iii.
3, 4, 5–15 ; iv. 24, 26, 27, 30 ; v. 7, 8, 9–14, 31 ; and as
each part of the human system is generally and, most
of its organic parts, specifically, *effected* by different
medicinal or poisonous or nutritious agencies, I. ii. 19 ;
iii. 5–15 ; and as it is well known that the whole or
diverse parts of the human organization are *effected* by

mental conditions, producing prostration or over-excite-
ment of the psychic organization itself, and of the visceral
parts connected with these spontaneities, these psytations, I. ii. 33, 34; ii. 7, 11, 14, 19; v. 12; and as it is
- known to the whole class of learned men in the medical
profession that medicinal agents, properly applied under
certain circumstances, produce certain general and special
and specific effects on what is by them called the Mind,
as intoxicating drinks, opium, hashish, canabis, &c.,
and that one branch of this profession apply specific
agents to specific characters of mental diseases, as they
are called, and this in the decided conviction, as gained
by experience as effecting them, as aconite, pulsatilla,
stramonium, calcarea, &c., &c., I. v. 25, and medicinal
agents and topical applications will be seen to be appropriate whenever the fact shall be distinctly grasped
and acted upon and the effective, the special remedial
agents shall be discovered and applied, that over-excited
organisms send their maligned influences along their
afferent nerves in to the Self and overpower its proper
normal action, and as these are excited or reduced
in action, so is the effect on the Self modified; and as
the effects of mental hallucinations — combined diseases
of the affectional, passional, and ideational functions —
in producing various effects on the human organization,
are seen in visions, swoons, wasting of flesh, hysterical
bleedings, unnatural appetites, as for human flesh among
the were-wolves of Silesia, as they were called, &c., &c.,
and in their nameless numbers of diversified forms and
perverted malignancies, as detailed by the numerous intelligent authors who have written upon manias; and
as the impressibility of human powers on the domestic
and other animals in their general and special training
is manifest; and as the influence of hypnotism and mes-

merism, and the recurring effects of degrading or ordinary
Imaginates on the whole nature of men who have not
been taught or have not the normal power of self-possession and moral conservation of their higher nature against
this *charlatanerie*, have become settled facts of science,
and these early vicious impressions are seen as originating in a fundamental law — in law-forces of action and
reactions; and as the influence of repeated opinions, involving the action of the spontaneous forces of the Self
and the manifest effects from a very ordinary eloquence
and false glossings of editorial non-responsibility upon
any subject and in any direction in which the popular
mind is prepared and impressed, show the reciprocating
correlations which subsist and which are so constantly
manifesting themselves between physical nature, somatic
organization, the instinctive, the psychic, and the spiritual
phenomenalizations, — the foundations are laid for affirm-
ing the immediate kinhood of the underlying forces
which thus reciprocate and interact.   In the multiplicity,
certainty, and definiteness of these facts, the suggestive
conclusion, yet for further elucidation, is affirmed that
the substances, subsistences, noumena — these antitheses
and causative origins of phenomena, I. iii. 20 — in their
secondary conditions and provisions for causations and
intercausations in the physical and moral economies of
the world, although scientifically different and separable
into distinct sciences, yet so interpenetrate and *effect*
each other, in a fundamental philosophy where all the
correlations and relations of nature and life are considered, that to a rigid analysis they constitute but one
philosophy, namely, Spirit, Πνευμα, in its broadest significance, actualizing itself through material forms in
spiritual processes from its certain definite underlying
ontologic forces.

3. While the Spirit of man modifies the material and
the psychic autonomy in which it is complexed, it is
modified in its manifestation by this material and psychic
complexus and by the reciprocating orgasmic forces interwoven therein and thus correlated to all the forces inwoven into nature and life, and by its own autopsic
spiritual independency and reaction within prescribed
limits — constituting its allowed circle. Beyond his
allowed circle he may not pass. It is not given to him
to institute an Insistent truth ; however reasonably or
fantastically he may apply that which is simple, attainable, and true in its elementary forms, and however va-
grantly and lawlessly he may form his ideations, he cannot,
in a system of correlations, escape from the Divine Ideas
in themselves and their correlations as woven around
him ; and whatever system of secular polity and disciplinary and active morality he may institute, it can only
be for a humanity founded in positive gradations and unfolding in a normal prolepsis ; — and whenever he passes
these bounds and prescribes an Immutable Morality for
the Almighty, he will reach those moral contradictories,
more irreconcilable and distressing than those which
rationalism finds in physical creation, and which involve
the moral sciences in doubt, if not in despair, and drench
the earth in blood for Revolutions to install the Rational
Principles of Nature, which forever escape in the ruin
and desolation and oppression of the people.

4. In the material forms, organic complexures, functional forces of their differentiate kinds, psychical processes, and evolutions and normal unfoldings of the Self,
and in the intuitions and ideations, and in their syntactic
correlations as cognizable by this Self, are the foundations
of all philosophizing and the consummation of all science
and progress. When facts are recognized as phenomena

of nature or mind, it is knowledge, and they are passed
over to the domain of Positive Science. Beyond this
ever-widening horizon of knowledge, but articulated on
it by the correlations to the transcendental synthesis, I.
v. 39, lies always and ever onward the region of philosophizing and the unfolding limits of Religious Thought,
yet based on those Intuitions and Ideations which embrace and complex and correlate the whole.

5. All elements of knowledge, as they are presented
in the Consciousness, are there subjective, either as sensations, instincts, psytations, imaginates, concepts, notions,
intuitates, or ideates, and as they are inwoven by their
correlations into actuation by the Intellectivity for the
gratification of the passional or affectional nature — and
below this they are the internuncial spontaneities of
instinct. Whatever may have been their original subjectivity in the Self, or the objectivity of the phenomena
which produced them, they must, as objectivities to the
Self, become enfolded in the reflex consciousness, and be
subjected to its scrutiny in its threefold light, and thus
become subjective-objects; that is, they must be received
into the clear consciousness and there be thrown into
objective position before they can be scrutinized and
made elementary to knowledge, philosophy, and religion.
Here they are enlifed, infecundated by the Self from its
animalistic, human, or spiritual subjectivities.

6. When so posited and become objects of meditation,
they are subjective, I. v. 20–22, and when its own
orgasmic motions in its animalistic impulsions and psychical psytations and its own self-conscious action, on and
through these, are subjected to ratiocinative processes,
the contemplation or analysis of these direct acts, passions, or affections, is the reflex action of the Self,
which can only occur upon the reproduction of sensation,

impulsion, or psytation as an Imaginate, and this through
the intervention and use of concepts, opinions, notions,
* intuitates, and ideates, at every step involving the correlations — the action and reaction of the forces inwoven
into nature and life. I. v. 25–35. A reflex act is the
Self reacting on the Imaginate restored to the consciousness by this re-enlifing process and in the ascent to more
abstract processes on the previous processes thus, in like
manner, reproduced. II. iii.–viii.

7. The only insistence which has clear and distinct
eternity, in its *formal* permanence, aside from the
Creative Forces, and which, in any final sense as a
predicate of language or rather as a category or essence
of the Intellective Power, of the divine Omniscience, is
the Insistent Truth. And yet this must be discriminated
against as not being ontologic, and as having only a
formal insistence ; otherwise this would be a return to
what has, by many, been imputed to Plato, namely, ideas
as eternal types or self-subsisting patterns. Being alone
is ontologic, and this is attained by the ideation and not
by the intuition ; but the insistence of this truth being
seen as an ever-accompanying law of concretion whenever any *factum* or created thing enters into existence in
time and space, it is to be recognized as the law of its
quantity. I. iii. 29. All else in nature and life is phenomenal ; yet the substrata of phenomena, by the objective act of creating, by the assigned immanence of the
eternal Creative Powers into it, may have an endless
duration, *e. g.* the souls and the spirits of men. When
the mind escapes from the theories of transformation and
emanation, I. iv. 8–10, and intuscepts creation, it can
only find immortality, eternal life, as a subjective identity, in the stabilitation of the eternal Will. I. i. 17, 18,
12 ; iv. 25.

8. When phenomena arise out of and from the Self, from the conscious exercise and normalative control or direction of some or any of its animalistic or psychical functions, they are *notated* with them in the organisms, and so are or may be made the objects of contemplation by the reflex action of the Self, — they are strictly subjectively-objective. I. iii. 17.

9. When the animalistic or human orgasmic psytations are not made the objects of reflex action, they are spontaneities; and pure simple spontaneities, *per se*, do not, nay, cannot, give reflective knowledge — human wisdom. It will be spontaneity to the end. Conscious self-direction and normalation is required for self-direction and self-culture in the cognition of these spontaneities and their *subjection* to the Proper Self.

10. When phenomena arise without the Proper Self and without the somatic and psychic complexus, that is, are presented from and through the agencies of the outer world, they are objectively phenomenal, but become subjective by the correlations which unite the whole in and to the Self. Let the notion of pure objectivity be taken as a somewhat standing over in objectivity from the Self, or some determinate act of the Self put over from the Self, and inwoven into the material contents of nature in form or action, and thus subsisting *sub modo* as from the determinating Self. Thus nature stands over from God. II. i.; III. iv.

11. The elements of pure Pneumatics, when speaking of the human solidaric Self, become consciously subjectively-objective, and are seen as ruling, directing, and modifying the orgasmic forces in the animalistic and the human organisms. I. i. 18–23; ii. 8, 12–15. 18, 19; iii. 9–15, 22–24, 29; iv. 9–12, 23–30; v 12–14, 21, 22–34. In these processes it has been seen that the autopsic

spiritual powers exercised by the conscious Self were the demonstrating Actuous power, the Intellective power, and the Affective power. These in their simplicity and in their complex combinations, as they arise out of or in the mental states, are seized and held and, as it were, are *re*-posited in the Self, and thus become subjective objects of discriminate cognition. I. v. 26. The want of this discriminate cognition leads to profound and continual error. This will be made apparent in the simple statements from the *Philosophicæ Institutiones* of· Tongeorgi, the most concise and comprehensive system of philosophizing yet submitted to the public mind. Lib. III. cap. I. art. II. § 277, 5°.

"*Volitio* est tendentia quædam in objectum *cognitum*. Nullus ergo potest exeri voluntatis actus, nisi cognitio aliquid voluntati præluceat, objectumque proponant.

" Pone nunc in subjecto non simplici volitionem. Erit in illo etiam cognitio. Dic, quæso, utrum cognitio et volitio pertineant ad diversas partes, an vero utraque in una eademque parte resideat.

" Si primum dicetur, absurdum dicitur, et quia est contra intimum sensum, et quia impossibilis volitio est, nisi in eo qui vult, sit, cognitio. Si alterum, non te extricabis, nisi subjectum in quo est cognitio et volitio in eo qui vult sit cognitio."

Again, *Id.*, cap. VI. art. I., he says, "Appetitus est tendentia in bonum apprehensum. Entia quæ cognitione pollent, in hoc differunt a rebus cognitione carentibus, quod non solum ab objectis quæ extra ipsa sunt, sed etiam ab objectis quæ in eorum apprehensione per sui speciem existunt, moveri possunt, si hæc ipsis convenientia sint. Et his motus appetitus.

" Duplex est in homine appetitus, sensitivus et intellectivus."

" Volition is a certain tendency towards a *known* object. Consequently no act of the Will can be exercised unless knowledge foreshadows something to the will and places before it the object. Now give volition to a subject, not simple. There will then be knowledge in it. Tell me then, I beg you, whether knowledge and volition belong to different (diverse) parts, or whether does not each reside in one and the same part. If the former be asserted, an absurdity is uttered because it is opposed to

17   .

the intimate sense, and because volition is impossible unless in him who wills there should likewise be knowledge. If the latter be asserted, you cannot escape unless the subject in which there is both knowledge and volition should have in it volition."

" The appetite is a tendency toward some good that is apprehended " (appreciated — desired ?). " Existences which possess knowledge differ from those wanting knowledge in this, that not only are they influenced by objects outside of themselves, but also by those which in their apprehension exist in their own proper species, if these be appropriate to them. And by these is the appetite moved. — In man the Appetite is twofold — sensitive and intellectual."

In both of the extracts, applying as they do to the identical Self, there is a tendency ; in both there is cognition ; and in both there is intellectivity for this cognition ; and in the last there is the same volition which is in the first, and therefore in the processes pursued by the author there are in one and the same thing, as a simple, elementary, or ontologic identity, cognition, volition, appetite, (appetency,) and tendency, if these two latter are not identical, and therefore in this homogeneous identity a tendency, a cognition, a volition, and an appetency. I. i. 8–10, 17–23 ; iii. 9 ; iv. Omitting any criticism which might be allowable on the change of *volitio* into *voluntas*, and this continued into *vult*, both of which may express *wish* as well as *will*, and referring to, I. i. 18, where will, counsel, and wish or desire are used convertibly in the Greek language as they are so frequently in the Latin and English, it is seen that the thought is as complete or defective if read, " because volition is impossible unless there be cognition in him who wills," or, " because wish, desire is impossible unless there is cognition in him which wishes, desires," or, " because to will is impossible unless there is desire (wish of some kind) in him who wills," and that each is imperfect as describing full normal actuations of the Self;

and that the three, the acts, the intellection, and wish or
desire, are requisites and facts of every complete normal
act of the Self.

The Self can have a certain tendency to a known object
only in virtue of a cognition more or less open and intelligent, and that to call forth this appetency in the Self,
this known object must have a use in some gratification
for this tendency ; and the self-conscious tendency to
any apprehended good, or escape,of apprehended evil, —
a negative form of good, — is appetency. Appetency and
gratification are as convertible as cause and effect ;
there can be no gratification except as there is appetency
and object and subject for the gratification. It must
be further seen that the appetency may exist without
the cognition, and to be called into action when the
cognition occurs, and the cognition may exist without
the appetency, and that in the human organization they
must be correlated to each other, and to the particular
objects in nature and life, to produce the phenomena
stated ; hence, in some degree, the diversity of natural
appetites and psychical tastes. And the Self may
know an object, may have an appetency for the object,
and yet not actuate for the object, until the cognitive
Self, from its other cognitions, shall determine for which
known or apprehended object *and* appetency it shall act.
Each product of these powers thus discretely complex in
the autopsic Self as it is generated, produced, or manifested, each in its province, is purely subjective and discrete in its genesis, whether the exciting cause is external or internal, but when reproduced from their notations
in the memory and made the object of these reflex
processes, is subjectively objective.

Pure Metaphysics as reaching to these correlate forces
in the Self, and to ontologic causations in Being, is there-

fore subjectively objective, — its facts being subjective facts presented under conditions of thought which must carry them up to the *a priori* law-forces which ruled the creation of existences and established their correlations in the grandeur of their vast and implicate system.

12. The Intelligibilities pervading the Self in these triplicate activities, and in the animal and vegetal orders in their intelligential functions, and in the former in their various instincts, (with but one general law for instinct, I. i. 30,) and matter as menstruums for recipiencies of more open and active forces, and these as wisely correlating forces, § 2, *ante*, necessitate in their systematic syntax the opinion, the notionalization, the ideation of ontologic forces coördinated in a base of Conscious Unity, the source and artificer of all in their impregnate and differentiate forces interacting in systematic correlations throughout their vast complexure. At the end of an ultimate analysis, on any of the filamentary connections of system, is God, the Creator, in the ormaic omnipotence and omniscience of his synthetic movements. I. v. 39, 41.

13. This gives the primal objectiv-faciency in creation, and the primordial division — Being — Existence. I. ii. 1.

14. Existence is matter in its differentiations, and differentiate forces, in all their forms of stabilitation, organization, and functionalization, and the Self with its complexus of triplicate coefficient powers in its threefold complexure of body, soul, and spirit.

15. Matter, each in its kind, is a somewhat — *a substans*, held together by its own inner constituting forces, I. i. 8–10, 12, 11, in which may inhere a congeries of qualities held together also by their own inhering correlating forces, yet capable of yielding their attractive tenacities and intensities to other qualifying, separating,

or more comprehensively combining forces, and thus throughout of adjusting modifications. Or, on the theory of Boscovich and Faraday, matter is forces *in situ*, or a complexure of forces held together by their cohering correlations, yet also with inhering contrapellences — powers of divulsion and repulsion. I. ii. 6; iii. 3, 7, &c. This is not given as the exact definition of that School, but it is the philosophic base of their doctrine. It will appear, as this system unfolds, that this stabilitation of forces is substantially correct, yet that there are created elements of matter, infinitesimals, as vehicles or menstruums for other differentiate forces, and all for the grand and majestic movements of the planets and star-systems in their places, the orderly working of organizations and the delicate tissues of the human brain which shall give discriminations of all to the autopsic man and from him react on nature and in life.

16. The Facts of nature — the finite — the concrete — physical phenomena always appear in serial orders, in causes and effect. Their consideration directs and leads the inquiring mind downward in their linked series through their actual and potential sequences to intermediary effects flowing from their intermediary causes to the final cause — when attainable and which is unattainable except in a moral system where the causative-end is enfolded in the beginning and rules the movement by a proleptic purpose — or, upward, to some causal idea coördinated in intelligent power and motive-love for their differentiate beginnings, their sustensive identities, and their actual and their phenomenal continuity. If the mind can overleap or rather connect and articulate in an unbroken series across the gulfs which seem to separate the beginnings — the genesis of species from their previous non-existence when new and differentiate

causes commenced their action, then the ascent back may
be to the first egg of night or germinative dot. I. iv.
5–8. But from the infinitesimals, falling into order
under the control and ordination of the ruling forces,
through to Autopsic Man, there is the harmonizing
system of correlations indicating the omniscient eye that
sees all, the governing power that rules all, and the wise
love that, in the prolepsis of his movement, cares for all,
and in the end this Love will, through his loving creation,
reciprocate the love in the beginning. I. i. 15.

17. The autopsic Self intervenes, or is superimposed,
upon all. Psychical spontaneities, and the autopsic Self
in its determinate movements, break up, alter, change,
direct, combine, and, with spontaneous or conscious purpose, use or abuse the serial orders of nature. This
conscious Self avails itself of that vast area of Intellectual and Moral Freedom presented in the philosophic
contingencies by which secondary causes may or may
not be brought into action, and the spontaneous or, in its
higher conscious life, the determinate Self is constantly
the agency by which this link between causes, more or
less remote from each other, are brought into contact
and causation for effects. These are conscious causations
constantly at work in multitudinous numbers, thus altering, changing, directing, combining, and using or abusing
the serial orders of nature. To what extent they thus
act upon, or may, in a general harmony of intelligent
and exalted motive-end, affect the general disposition of
nature, philosophy or science or religion, we have not
seriously inquired, nor yet proposed as a definite end of
inquiry, although nature in many localities and in some
of its species is evidently improved by the sporadic or
spontaneous, rather than by any determinate movements
of civilization for a determinate system of improvement,

while the fact of stationary or retrograde condition of man everywhere on the earth is actually typified in the condition of the country which he inhabits. This may be illustrated by the condition of Judea and Galilee in "the mantle of barrenness with which the demon of Islam has covered it," in the improvement of the cold and inhospitable forests of Germany, in the greatness and in the improvement, wealth, and comfort of the Netherlands, in the valley of the Nile, in the plains of India and the wastes of Africa, and in daily life in the dilapidated fields and home of the drunkard, in the viciousness and contaminations produced by the expenditures of the profligate and in the contracted poverty of the miser, and in the world everywhere, where man touches it in his rich penury, in his lusts, voluptuousness or pride, or desolating power. The psychical phenomena, thus closely interblended with physical nature, more or less typify these serial orders, though seemingly more irregular, erratic, and incalculable from the automatic acts of insanity, monomania, or unbalanced organization to the conscious acts of deliberate autopsy, and throughout they are subject to the calculation of opinion in the fact that all nature and life are bound together in a system of correlations. These spiritual causations in man, in the selection of his ends of action, in the intentional breaking up, altering, changing, directing, and using or abusing the serial orders of nature and the moral forces in life, and shaping, moulding, and constructing from them means and instrumentalities to the selected ends, manifest and vindicate their own independent and autopsic character, yet always limited to means and ends within the range of the appointed prolepsis and its own relations in its time and place.

18. This autopsic Self, in its higher foundational —

solidaric action, is that unity in triplicity which possesses
the consciousness of sensations, external and internal,
and of the direct acts and reflex consciousness of objectifying, intellectualizing, and loving. I. i. 4, 17, 34. Its
sensations must be as the natural world is correlated to
it in appropriate organisms for perceiving its quantities
and qualities. And these must be as the place of our
world in its star-system and as the place on this planet.
I. iii. 1. In such a theatre and in such an organization
it must ever be the slave or the master of the objectivities with which it is surrounded, or it must aspire to an
End above these objectivities; but it can do so only on
condition of passing through these articulations around
itself. At any point in the movement in which it stops,
there its progress and ascent ceases. Well is it for him
to whom change, vicissitude, sorrows, and deaths come as
messengers loosening his fetters of the lower conditions
and bidding him to aspire. His end is not found in the
animalistic or human loves of gratification of this world.

19. Consciousness — Self-consciousness is the foundational fact of autopsic life, whether this be found in
the knowledge and love and actuation of goodness or
guilt. It is the primary fact and foundational cognition
for any fundamental philosophizing. Its spontaneity is
impredicable to philosophy. It cannot be thought, and
any philosophy of life be preserved. And in Omniscience all spontaneity is of necessity excluded. Omniscient Consciousness is universal cognition; and circumscribed consciousness is limited cognition, and as these
limits are enlarged, man ascends towards the universal
cognition; so in the love and so in actuation. So Man
learns and aspires. Again, the End is seen in the beginning. This consciousness in the Self is the eye, if
not " the light of our seeing." It is the witness of feel-

ing, of actuation and loving in the dark labyrinth of the organized Self to be brought to light in its great fulness and ever onward realizations of its conscious intusceptions. It is the self-cognition of the abiding energizing of the Self in all its normalation. It is the only attesting witness of every such exercise or act of mind, and through all such mental phenomena it cognizes its own identity, which it never loses, save in insanity, pure instinct, sleep, unconscious revery, spontaneity of passion or affection, or in some lesion of the brain which breaks up the intercommunication of the Self with the outer world, or so affects the organism that the Self has no control over it, and the organism so affected ceases to be the organ, so far, of the Self in its intercommunications. In the fact of its control, in the normal conditions of the organisms, is seen the independency of this Self, — and in the automatic action of the organisms their separate and organic use and action are made apparent.

20. It has been shown, I. ii. 7, 8 ; iii. 5–15 ; v. 2, 3, 7–14, and the influence of these facts will more fully appear hereafter, II. iv.–viii., that all sensations are effects upon an outer organ or upon an inner system of afferent nerves which carry the registration of the quantities and qualities of objects or the modifications of some of the internal senses, as hunger, thirst, erotic desire, &c., to the place of perception, the *locum tenens* of the Self. Around this place, this *locum tenens* of the Self, are placed those organisms which assimilate man in so much of his human passions and affections and capacities with the nobler forms of the animal races, — the cunning of the fox, the wonderful constructiveness of the bee, which performs blindly-wisely that problem which required the highest mathematical skill of four of the ablest men of Europe to solve and determine, the singing of the

mocking-bird, the ferocity of the tiger, the accuracy of
the little chætodon which, from its denser medium in the
water, can shoot a drop and strike a fly in its rarer
medium and make it its prey, the boldness of the lion,
the sagacity of the elephant and the dog, and the mater-
nal instinct of all animals, and from its superior and
more or less independent position the Self must use these
organisms in its consciously normalated life. The fox
is not cunning because it self-consciously determines to
be so, but when the impinging of its hunger repercusses
to this organism, the whole animal is reflexly put into
motion to accomplish its gratification and its cunning acts
instinctively, as ideational, passional, or affectional manias
act in man. But in the life of self-conscious normalation
the Self must say to this cunning, Shall it be used for fraud
or to attain a proper end by proper means? it must say
to its bird-singing capacity, Shall it be æsthetically cultivated for the pleasure and adulation of men, or shall it
be for praise and joyousness and gratitude, or for the sor-
rowful songs of wrong and exile where they "sit by the
waters and weep"? or to that ferocity, terrible in all
animals, but more terrible in man with his murderous
instrumentalities, shall it be used in deeds of horror and.
shame, or shall it too but lend its forces, abstracted from
it as the fish became blind in the cave, to the moral
powers to sustain them in the otherwise unequal conflict
with Force and Fraud? And so from its independency
and supreme height the autopsic Self surveys the powers
of nature and life and ascends to higher heights. I. i. 1;
iii. 29; v. 12, 14, 15.

21. Phenomena are the manifestations in and through
the sense-bearers or some appropriate organism, to the
Self, of the law-forces operating in and from material
substances or in and from animalistic and psychical or-

gasms and their organs, and outwardly from the autopsic Self and other selves. I. v. 3, 4.

22. Cognition, in its most comprehensive meaning, is the simple act of mental beholding, cognizing by the accurate perception by the conscious Self. The simple fact of cognition can be negatively raised by the scriptural expression, " having eyes, they see not," having the organism and the cognitive power, they do not cognize. In beholding in or through the senses, the cognition is at first slow and progressive, the capacity of cognizing clearly increasing with the expanding consciousness — in the expansiveness of the cognitive power breaking through, as it were, and enlarging its organisms, — not in the intensiveness of the pursuit which tends to blur the consciousness. When this intensification is subdued by reflective habits, it makes the cognitions more sharp and incisive, more in relief, while the intensification of the passions and the affections, as has been so frequently shown, gives a constant tendency to automacy in some of its many forms. J. iii. 9. In beholding in or through the senses, the cognition is direct, and in some instances the single cognition, so made, is the only test which the Self possesses for the verity of the cognition, while in other instances the Self has the means of verification in some other or others of the senses. There is the same certainty in the conscious cognition and appreciation of the Registrations brought by one sense that there is by another ; and although the Self may, at times, appropriate the sensation brought to some other than the true object, yet in the sum of all the verifications given by the organisms, which apply with so many correlations to the objects of nature, there is a general and satisfactory verification of knowledge. In these several processes the Self is in the conscious state of a clear beholding,

applying the seeing by the eye as a metaphor to the other sensations and to the psytations, and this beholding is cognition. To see is to know by the eye; to hear is to know by the ear; to taste is to know by this sense; to hunger is to know by the stomach; to feel anger, wrath, love, charity, is to know by the palpable registrations of their effects on their various correlate viscera — as in bowels of mercy, shame on the cheek, sinking of the heart in distress, firing of the chest in indignation, fear in its effect on the system generally, and in much fear in relaxation of the sphincter muscle, &c. &c. I. iii. 9. Here, in the region of the Sense, discriminations become necessary to, and are a law and fact of, the mind; and are only the cognition of the Self limited in its cognition by the limited nature of the object or subject-object which gives the fact, and by the limited nature of the organism which conveys the fact in to the Self. The Self retains the facts, and recognizes their kind, their similarity and difference, simply in virtue of its cognitive, its intellective power. So through the entire region of its cognitions. I. v. 8–14.

23. There is an external relation which all objects in nature bear to each other in time and space; they are situated here or there; they are up or down relatively; they float in the air, or creep or crawl, &c.; they were, or are, or will be; and these relations in their primary cognitions are of the sense knowledge and they give time and space. But there are correlations of things in themselves and of things to each other which are seemingly not given in the sense, but such correlations are seen to unfold themselves through the Sense to the Self when the diremptive analysis is applied, and they are taken apart piece by piece, constitutive element by element, and the correlations conjoined in the cognizing

synthesis of the Self. I. v. 17–20. The conjoining is given in this actual cognition of the actual correlations. The cognition of relations and of correlations is a simple beholding — cognizing by the Self. When all these relations and correlations are seen as the web of a great complexure surrounding the Self, and touching and interweaving with its whole nature of knowing, doing, and loving, as these unfold in the Self, they are bound together in the very filaments of this ever growing and expanding web. The simplicity of the process would be more apparent, did we not so constantly get the mere results and not the processes, as " Kant, reflecting on the differences among the planets or rather among the stars revolving around the sun, and having discovered that these differences betrayed a uniform progress and proportion," conjectured the existence of Uranus — he but widened and applied the correlations already known and attained in many centuries, by the same process, to a point in space which needed this new correlation to make the system of correlations perfect. Franklin surmised that the electrical spark and lightning were identical, yet what facts had he collected, cognized, before he bound them together, and in the cognition of the identity of their phenomena pronounced them the same. So Newton, standing on the vantage-ground of all previous approaches to his great cognition, saw in the fall of an apple the correlating force and law of gravitation.

24. Sir Wm. Hamilton speaks of the Noetic Faculty, (the later German, French, and English metaphysicians and psychologists call it the Vernunft, or by an equivalent term, as contradistinguished against the Verstand, now affirmed as the mere organic understanding of animals and man,) and claims the former as being the *locus principiorum,* the place or faculty of the Common Sense of

the Scottish School of thinkers on this subject, and as
giving from itself, from its faculty, *locus principiorum*,
place of principles, certain primary cognitions — not as
cognitions but as of original inherent knowledge. I. i. 35.
It is not the purpose to discuss the number, character, or
special uses of the organisms, I. i. 17–23 ; iii. 7, 13, &c.,
through which the Self manifests itself to the outer world,
satisfied that it can only so manifest itself by and through
special organisms, and that it is in this way alone it can
be apprehended by other selves, and that it can in the
same way, only, gather through organisms all that the
outer world has to give or convey to it ; yet it must be
seen that the cognitive Self, as a pure intellective agent,
as a simple intellectivity, is a something, one and indi-
visible in itself, while it must gather its cognitions through
organisms, and thus " be renewed in knowledge." If the
intellectivity, the cognitive power, is other than a simple
unity manifesting itself through organic functions, it can
have no sovereign authority for direction of conduct, for
election of ends, for arranging in harmonies of unitary
system the vast accumulations of its various cognitions,
passions, and affections. But all these, the elements of its
cognition, can only be received from a world of organized
correlations through organisms adapted to their transmission ; hence in the history of the human mind there is a
progress in cognitions ; and hence it is that, as a higher or
lower form of civilization or of religion is written upon
these organisms by education and position in the prolepsis,
the whole mental characteristics of the race so approx-
imately correspond to the impressment made. The only
intelligible argument for the *being* of God is the unity of
design which binds all nature and life in a system of
relations and correlations. It is the unity of the Omniscient Intellectivity of God. Omniscience is universal

cognition. The unity of the Intellectivity of man while
it is manifest in the receptivity of its cognitions of the
facta of nature and life and their correlations and of its
own conscious demonstrations and of its arrangement of
all these elements of cognition into intelligent system,
so these can only be attained and normalated in virtue
of its own unity, by which it reconstructs the *facta* and
restores the correlations to their appropriate adjustments.
The cognitive Intellectivity which perceives, knows *this*,
is the same intellectivity which knows *that*; the same
intellectivity which knows this and that, knows, cognizes them in their differences and similarities, and the
same intellectivity joins them in their correlations and
perceives them in their repulsions; and this again is but
analysis and synthesis and redaction, yet is but *the* one
cognitive Intellectivity. As the organism is imperfect or
inauspicious, the Self may fail in one or both. It may
not gather the facts properly or fully, and it may not
arrange them in the disorder or want of normalation or
normalative condition in the organisms as its instrumentalities. It mobilizes through its organic instrumentalities.
In any other view the Intellectivity is but as the natural
life, the result of organization, and is not an indestructible
something unfolding its power as it perfects its instrumentalities, or as they are perfected in the causations
which are at work in the philosophical contingencies in
the web of complexure in which it is placed. I. i. 13 ;
ii. 3. And the argument which shall make the Intellectivity the result of organization will establish the proposition that the wisdom of the universe is the result of
organization, not that a Unitary Intellectivity organized
it. The facts of life are constant, and there is a certain
uniformity in the facts which show that the organisms
are constantly improving or deteriorating in the wear and

tear of life, and in the philosophical, religious, moral, governmental, and geotic influences which surround the Self in their actions and reactions on each other. If the Intellectivity is the result of organization, or something which in itself is alterable by these causes, then it is a pure *accidence* of these causations, and so is not the simple unitary correlate in man or coördinate in God. The Cognitive Self is one knowing agent which cognizes and affirms its cognition in words or deeds simply in virtue of its cognitive Intellectivity from elements obtained in percepts, imaginates, judgments, concepts, notions, opinions, intuitions, ideations, differing in individuals in the clearness of the cognitions and their arrangement into system in some definite relation to the perfectness or imperfectness, sanity or insanity, congenital or superinduced automacy or lesion of the organisms. II. ii.

25. All cognitions are simple affirmative judgments of the Self. Each cognition is a judgment; it is so or it is not so; a man is seen at a distance, and the object is judged to be a man; as he approaches and the cognition becomes more defined, he is cognized as a man; as he approaches more closely, he is judged to be this man or that, and in the certainty of cognition of the knowledge it is a certain man. Revise the process: the Self approaches truth by uncertain and ill-defined or imperfect cognitions, but as the Self, in its intellectivity in pure intellections and in its triplicate intusceptive processes, in cognizing nature and life, approaches, in its gradual cognitions of objects in themselves, their differences, similitudes, and correlations and opposing identities, I. iii. 18; iv. 3, 4; v. 6, 8, 19, each step is an affirmative cognition, and in a fundamental philosophy it is the cognition by the intellective Self of its own intellections, actuosity, or love, as they apply to objectivities as moral or physical causations.

§ 1, *ante.* These cognitive judgments are affirmations made in the analysis of observing parts and elements; in synthesis in putting these parts together; and in redaction in seeing — apprehending them in their forms.

a. An analytic judgment is a diremptive or discriminative affirmative cognition. In a lengthened process it is an affirmation of successive affirmatives. In the elimination of truth from error it is the process of putting successive affirmative cognitions in their correlative order so as to exclude the error, or it is the affirmative negation of the erroneous affirmative implying in the negator a previous affirmative cognition of something known — cognized by him. Thus from the definition of a point, a line, angle, to the demonstration of the most abstruse proposition, it is a succession of affirmative cognitions obtained in previous analysis or separate cognitions, and bound together in the process of the demonstration by the synthetic cognition discerning the parts in successive processes and in the whole. In physical natures parts are cognized and joined in greater parts, in fragmentary systems, and in a final system. In moral life the cognitions come through the animalistic and the human orgasmic powers and passion, affection, lust, cupidity, and voluptuousness, and the cognition which leads to action must partake of the elements of cognition and the affinitive forces which are brought into action. The process of the divine prolepsis interposes in its ruling ordinations, and in the griefs, sorrows, and disappointments of life breaks down these passional and affective influences ; and as this is done, the intellectivity *may* become deobscurated, according to its relations in time and place in the movement, and be correlated by progress to a higher range of life.

b. A synthetic judgment is the affirmation of accordant

18

correlations or many or successive according correlations
in nature and life, or in the highest cognitions it is of the
coefficiencies in the primal Causative Forces. The synthesis which binds nature into a mere Intellectual System
is from the native elemental capacity of the intellectivity,
but the synthesis which binds nature into a Moral System
is from the conjoint action of the triplicate powers in the
Self. And thus it will be seen that as man is in a low
and degraded condition of organization, his God and his
system of the world will have an exact correspondence;
and as he takes counsel from the tiger and the fox and
the serpent within him, so will be his God and his means
and mode of executing his opinions of the wishes of this
Deity fashioned so like himself. But when he ascends
to the simplicity and grandeur of his true spiritual nature,
and reverses this action and impresses on these organic
powers the sanctification of a depurated love. I. v. 33–
35, the fruit of his spirit will be love which shall suffer
for sin in others and repeat not — as at Jerusalem on the
Cross, at Ephesus among the wild beasts, at Rome in the
bloody persecutions, in own obedience to Truth. § 20.

c. The redactive cognition is the presentation, in word,
symbol, or system, in some form, of all the previous processes of cognition, or of detailed parts or of any whole
and of the whole. As Deity when he creates can only
create in Form. L i. 23, 24, most of nature and life is
given to the Self in forms, and these he must break and
disintegrate to obtain the analysis by which the Great
Worker descended from his synthesis into the details of
creation. L v. 29. At the end of any process, in virtue
of the analysis which gives the cognition of details and in
the synthesis which restores their union and finds their
correlations. Form supervenes as the limit of action of all
the forces or processes at work. At the end of pure in-

tellective processes it is *pure* form, as in geometry ; at the end of all the processes by which the Self would grasp nature and life by its threefold intusceptiveness it *must* be God in his coefficient powers. I. iv. 28, 31.

26. A negative judgment or cognition is, therefore, only the denial of an actual or presumed affirmative cognition. It is the negative pole of the affirmative, and therefore in the processes of the disenvelopment of the true may be a negative pregnant containing the affirmative truth, or it may be, simply, the negation of an error ; but this can be only on the ground of some actual cognition containing in it the contradiction of the error — the orderly arrangement of correlations showing the negation is true.

27. These views exclude, except as a mere rationalizing form of routine, the logical formula of " Contradiction and Excluded Middle," as being a mere logical negation of one of two affirmatives, one of which the cognizing Self knows, perceives to be untrue, or one of which it knows, cognizes to be true. In the proposition that a thing cannot *be* and *be-not* at one and the same time, there is only an affirmative that the thing is or that it is not, — in the latter instance implying an affirmation that is cognized ; and it is thus only the elimination of the formal contradiction between an affirmative and a negative affirmative, leaving the proposition still to be determined by the simple affirmative of the cognizing Self, and which is the final affirmative in all processes of cognition.

28. Philosophy is, therefore, the cognition of facts (physical, psychical, and spiritual, in which the correlations in their syntax are as much facts as the more concrete forms of *facta*) in affirmative explications, and as those facts and their correlations are presented in the modes and in the manner set forth. I. i. 31–33 ; iv. 28–31 ; v. 1–16.

29. These processes again return to and philosophically
necessitate three inquiries, which have been hereinbefore
broadly stated and illustrated in general views, but which
will be demonstrated in the positive cognition of the facts
and correlations of nature and life.

a. What Causative Power extruded, produced, projected into objective immanence, matter and soul, so wholly
and phenomenally unlike, if not antithetical in their phenomenalizations, and yet in the latter so like the forces
which move and rule and subordinate matter in the dynamics, plasticities, and autonomies of nature, and superimposed upon them the Autopsic Spirit, which, within
its assigned and allowed circle, moves, rules, and subordinates both under laws of moral correlations, yet in virtue
of the very correlations by which it moves and rules and
subordinates, and is reacted upon by them. I. v. 31, 32.

b. What Causative Intelligence, Wisdom, Intellectivity,
gave them their modality in detail, their formal redaction
in their existences, as species, &c., and integrated, inwove,
in them, from his preëxisting Ideas and coördinate forces,
the forces differentiated for their functions in dynamics,
plasticities, autonomies, instincts, psychical orgasms, and
the endowment of the Autopsic Self and their wisely
adjusted correlations, for their permanency, phenomenal
development, and action and interaction ; some with more
or less stabilitation, others with plastic forces of action
and reaction, others with autonomic forces of development, growth, and decay, and reproduction, others with
instincts fiercely blind and intractable or docile and ductile, others as fierce, as docile and ductile, operating in
like manner in the human brain and on the viscera, and
the other and the last possessing or capable of unfolding
the conscious power to modify and rule all these, and from
his transcendental ideations and intuitions to apprehend

a system of moral life for itself, and to exhume, from its degradation within him, his holy love, and from these ideations unfold in it the Sense of Responsibility, and, in the mazes of the labyrinth of his life, to feel and perceive the necessity of commands and obedience for his lower life, and in this very obedience to commanded duties, sinew the soul by the integration of higher assimilated and depurated forces, which, in the movement of the ages, • mould brain and skull and face and form. As into the blacksmith's arm, the Self, by its determinate actuation, throws its vitalizing influences, so the Self notates, vitalizes the brain and the respondent organisms of its psychical powers. I. v. 25 ; II. iv. The Self thus builds up around itself its correspondent form and its respondent destiny. As it sows, so shall it reap. I. v. 31, 32.

c. What was the *impetus animi*, the coördinate orma, speaking as Aristotle intimated when he said, " the principle of reason is not Reason, but something better," which Tongeorgi but yesterday implicitly sought when he said that there was an intellective appetency — *appetitus intellectivus* — which the yearnings of all noble philosophies indicate in their desire for and in the pursuit of the Good, the Summum Bonum, and which forever fails in mere intellectual systems and human pursuits. What was this impulse, holiest of holiest, which induced their extrusion, and placed all things in time and space, in their concrete relations and conditions of development and normalation in the complexity of their surrounding and supertending correlations, and inwove a love in the foundational movement of all life, weaving the web of existence with threads of golden light, showing the Causative Love in the beginning leading to and designating a reciprocating Love in the Causative End. The Prolepsis moves onward. I. v. 31–34.

30. Following in nature the series of effects and causes,
upward, through its linked processes, and over the broken
and disrupted chasms of old series and new beginnings,
where beginnings of new species differing by wide forms
of organisms and differentiate functionalizations of forces,
I. ii. 12, where the beginnings of these new differentiate
species are seen, the Self is conducted to Efficient Cause,
binding nature fast in its fate of cause and effect, and
governing nature by its superintendent determinations in
its new creations, and retaining scope for special providences by the adjustment of correlations dependent on
the philosophical contingencies. Following them downward, holding the clue of intelligible causes and effects,
.thus enlifed with the depurated powers of the Self, in one
hand, and the proleptic light of causes working to higher
ends and pointing to the future in the perfectibilities of
existence which illuminated the past, and radiates, although through red and sulphury clouds, on the present,
the Self comes, in its intense serene, to the investigation
and some knowledge of the Initiate — the Efficient, the
secondary, intermediary, and the Final Cause, — being in
one whole the Final Cause. Efficient — Initiate Cause
and the Final Cause, though separated by the whole intervening series of causes and effects and autopsic powers,
acting within their allowed circle, must be coincident and
consentaneous; for the idea realized, actualized by the
creation in formulations from the deific intellectivity, and,
infecundated and enlifed by the ormaic movement which
in his Love arranged the prolepsis, organized creation,
imposed dynamic and plastic forces, and autonomic forces
of development, psychical forces in the impulsions and
solicitations on which to hinge, unfold, and perfect the sense
of responsibility in the autopsic Self into a free obedience
to holiness, and will close the period in accordance with

the primal idea. Rigid stabilitations of the thick-ribbed earth, giving oceans, continents, impassable mountains, and deploying lines for the children of the races writing and completing in the great drama of the world the sublime Unity of its thought, by which all are bound together· in an accordant and irrefragable system in which, while ultroneous natures shape their own destiny or ascend to Truth and Love, reveal and actualize the unity of the Divine Idea. I. i. 41–43. This idea, in its threefold constituencies entering into the constitution of nature and life, and running throughout the many-folded series, is the Prolepsis which binds the objective, immanent whole into one unitary movement.

31. No one Self is wholly intellective ; no one wholly under the dominion of any˙ one love or pursuit of gratification ; no one is wholly actuous ; nor indeed can be and be man — or Deity. But as either predominates, the character will correspond. These elements are mingled in each self, and this with more or less complexity of the animalistic and human organisms, and their capacities for assimilating their special orgasmic forces. I. iii. 9, 15. The great diversities of races, tribes, and individuals, in the vast successions of life, in their diverse localities, give the actual conditions of human gradations from where the distinction between the animal and the man is scarcely perceptible except in outward form, through, up to the ministering agencies of the holy Love ; and they give throughout the moral necessity for discipline, instruction, and education to receive or perceive the higher ministrations unfolding still beyond, and to secure further progress by obedience to the command, which, slowly, at first dimly, but gradually and surely, leads to higher cognitions and purer love, — and below this is the actual and positive necessity of sub-

mission superimposed for the order and peace of society,
and as the preparation for instruction and education in
the centuries.. From these diversities results the necessity for government, the institutions of society, and the
higher moral necessity for the teachers of truth and
holiness in the Lord. But all governments are incom-
petent to attain order and peace, as the history of the
centuries plainly attest; nor does ecclesiasticism give the
conciliation; and both united frequently, almost always,
in the complexity of the causes at work in human life,
gives the intensifications of fanaticism to the secular
power, and the secular power lends the sword of decima-
tion or extermination to ecclesiasticism; and thus united
they solidify in centralism and a rigid despotism; while
the unnormalated conduct of men, without that government which every community ought to furnish for itself
to keep order and peace in its own borders and punish
Fraud and prevent Force, will end in confusion and
anarchy. Such are the paradoxes with which humanity
must deal. If government oppresses, man will rebel;
if ecclesiasticism, of one kind, breaks the idols and *pro-
fanes* the temples where others offer sacrifices of folly and
blood, the knowledge and the love of which are inwoven
by persistent registrations in their hearts and brains,
and affectional intellective organisms, I. v. 26, the direst
vindictiveness is aroused to atrocious actuation; yet if
to thee has been bestowed the terrible but lovely gift to
know God in the genial, inspiring, and inexorable Moral
Logic, teach thou those lessons of purity and love which
would lighten the labors of all in the self-normalation
of a holier life, though ambition and pride and cupidity
and lust shall pursue thee and baptize the truth and the
life in sorrows and blood, — for the Sufferings of Love,
in less than three centuries, amid the most horrid in-

humanities of secular power intensified by the fanaticism of the ancient ecclesiasticism, through ten persecutions which filled the world with fire and slaughter, made thirty millions of converts to the doctrines of Peace and Love.

32. Man, in his organization, is animalistic and human, and he is Spirit. To amplify and not to degrade any conception of his nature and destiny, but expose the conflicts he must encounter and the victories he must achieve ; the swine, in his organization, closely resembles man in nervous tissues, muscle, and viscera ; they have appetites, passions, and affections in common and alike. Here the laws of the accumulation and expenditure of forces intervene. Contemplate the volume of forces taken up into their respective systems from the air, the food, and water. I. iii. 5–15. The expenditures of these forces are palpably appreciable in the motions and exertions of their bodies, and these must be seen as flowing in their natural currents to supply the exhaustion occasioned by the exertion, and to strengthen and increase the organism exerted, — the blacksmith's arm. The expenditure of forces is none the less appreciable in the action of some of the more intense gratifications inwoven in the organisms, as in sexual intercourse, and the law of natural supply in animal and man is the same. And that the effects on the respective systems in altering the organisms is the same, is equally apparent. When alteration is the result of moral control in man, it is seen producing effects on his system greatly different from those by natural excision, while man can become more brute than the brute by the infusion of his conscious powers into the animalistic orgasms. Ascending thus throughout the animal, and the animalistic, and the human organizations, it is appreciable that the organisms through which

the Self works out into actual life are supplied by forces
from the assimilations of the plasticities prepared in the
processes of nutrition, circulation, and assimilation, and
as these fail, their demonstrations fail.   Up to this point
it is seen that in the ganglionic centres where these respective forces are prepared for their specific offices or
functions there is an intelligential power inwoven which
acts ganglionically in its prescribed order to produce the
specific action of that part of the natural body ; and the
brain is the centre of ganglions preparing the psychical
forces. I. iii. 5–15.   But here it is that the conscious
Autopsy comes in with its determinations of conduct
from a higher and other source to react intelligently on
these functionalized forces and modify, direct and control,
and use or abuse them.   In this ascent it must be seen
that the animal, impelled to action by his orgasmic forces,
cannot rise higher in conduct than the driving forces of
his nature, and that man, in so far as he is governed by
his animalistic nature, is but animal, however he may
cover up these indulgences in the embroidered mantle of
his æsthetic forms of culture.   In man he is but man by
the human laws and forces of his mere human movement
in such his higher organization.   This view gives the
facts and the laws of the two lives.   Directly opposed
to these is the Spiritual Life, yet with power to mould
and use these natural organs in a sanctification of means
to ends, and, by denuding them of their animalistic and
human tendencies, to build a higher organization. I. iii. 15 ;
ante, § 2.   As animal he is but animal; as man he is but
man ; and as man he may reason and act from the animalistic impulsions, and as he gratifies them he increases
their fatal tendencies ; so as man he can only reach to
the heights of man with all their chances, changes, and
disappointments, and throughout his logic will be as are

these passions and appetites. I. iii. 29 ; v. 26. In the full possession of his spiritual nature and destiny before him, he aspires to his end in Love for Love in the End — which is God. So he must aspire by his intellective and moral freedom through forms, ideas, commands, laws, principles, motives, sentiment, and grace as provided in all the Revelations and Inspirations for his ascent.— and this in virtue of his Intellectual and Moral Freedom.

The Imperfect aspires to the Perfect, and so long as these clinging impurities shall stain and shall have stained by their direct ingoing action the purity of his spiritual nature, he cannot stand in the Solemn Presence, and therefore there must be, in some more or less inscrutable way, a long line of depurating efficiencies cleansing his nature, or a conciliation by some satisfactory mediation, or by the combination of both.

33. Intellectual Freedom is a choice of means, morally indifferent in themselves, for the attainment of an end, which end is a moral indifference.

34. Moral Freedom is a choice of means to an end or a choice of ends, in which means and end must be selected in conformity to Principle inwoven in Law and Command, or the command and the end be pursued in obedience to the law or the command, on condition of penalty ; — the law and the command, the means and the end, involving the actualization of a principle in a spirit of obedience in love.

35. Command, as a subjective noun, is the personal right and power of governing intelligent personalities with exclusive authority. It implies moral government in the direction and control of the powers of the governed. It implies power to enforce and regard for the commanded. It includes intelligence to perceive the moral end and define the just mode of the government ; and it

supposes a moral defectiveness which can be improved by punitive *causes*. The right to command, govern, can only intelligibly be found in the end, and the end must be proleptically contained in the command; that is, the command must foresee the end and shape the command so as to conduce to the end, and the end can only properly correlate to the command in the temporal good of a temporal government and the eternal good of an eternal government. This correlation not only implies intelligence for the adaptation of means to an end, but it implies regard, care, *love* for the governed; an interest in their welfare that the means shall be adapted to the end, and the end shall be the welfare of the governed for the reciprocation of this regard, care, love. This intelligence to devise — select means — the power to enforce their observance in the demonstrations of effects, and this love of the final end for the good of the governed, must be found in an Unit or Unitary *complexus*, and must have the three elements of a Personality herein included and set forth; and this, whether it is an aggregation or corporation of many persons collecting these elements into the governing centre, or whether, in an eternal government, this Unitary Centre is the coördination of Infinite Power, Absolute Intelligence, and Unfailing Love in one Supreme God.

36. Command, as an objective noun, is a rule and a principle of action prescribed for a moral inferior; imposed by one having the exclusive right and power to govern; which the inferior should in the light of a sufficient and enlightened intellectivity obey, for his own welfare and obedient love to him who commands; which he must obey under penalty of displeasure of him who commands to be manifested in special and appropriate discipline; yet the principle inset in the command may

be disregarded, and the command disobeyed by the person commanded, on his own choice of self-gratification in what is prohibited or incurrence of discipline and penalty. It implies in the person commanded intelligence to comprehend the command, and more or less of power to obey; to do the positive, to keep the negative commands, and to reciprocate the regard, care, love, which instituted the command. It virtually evokes or challenges a love for the sake of the Wisdom, Power, and Love employed in the government of the commanded. Command, therefore, differs from law, in its proper, philosophic meaning or rather law-force in this, that law-force by its inherent force inwoven in the means (*e. g.* instinct in animals, gravity in the falling stone) secures its end; — the command may or may not be obeyed.

Command, as aforesaid, may also be applied to persons of inferior intelligence with or without capacity to learn, but who may be compelled to submissiveness for the sake of the order of the whole, and in those who can learn as the discipline for further instruction and education.

It therefore follows that government cannot command any temporal policy which is not for the temporal welfare of the whole, and that it cannot command obedience to a moral wrong.

37. There is no word in the English language, and in the correspondent terms of all languages, more frequently used without a vital thought to enlife it than the term Law, νομος, *lex, &c.* When the term Law is applied to a command to an intelligent and responsible inferior commanding some act or line of conduct to be observed or omitted, and which he must perform or incur a penalty, this is Law. But when the commanded, in

virtue of the law, observes the injunction, or violates the
law, and the penalty is inflicted, then it is seen that the
Law, in and of itself, does not execute itself: — in the
first instance, the agent commanded executes the law by
an intelligence and a power personal in himself respondent
to the intelligence which gave the Law; in the latter instance, in the infliction of the penalty by the law-giver,
there is a new element brought into exercise, namely, force,
in some form, for the infliction of the penalty. When Law
is applied to a conscious agent as a rule of conduct
which he must responsibly obey, it is a far different thing
from law applied to unconscious matter; and if applied
in both instances in the same sense, one of them must
be absurd. In this latter sense it is almost, if not altogether, used in current philosophies and theologies. In
I. iv. 3 and 4, the process for arriving at the true meaning of Law and Law-forces is brought into view. Law,
therefore, in the latter sense, is but a cognition of the how,
of the mode of forces acting in matter, and does not at all,
in correct modes of thinking or expression, give the ideation or valid conception of *powers above* matter inweaving
in it regulative and functionalizing forces which it must
obey, and by which it executes the movements of nature
and life. Law, as a precedent idea for the inweavement, functionalizations, and stabilitating and moving
efficiencies of nature, is only the precedent intellective
form which is thereafter to be filled and enlifed with a
content of positive forces. It is but *idea* — the Divine
Ideas. I. i. 37, 36, 35; iv. 15. Starting from this point,
it must be seen that the antetypal idea and the stabilitation and functionalization of forces are correspondent
one to the other, and that the latter is representative of
the former in the concretion, actualization, and actuation
of the primordial forces which created. Law, in this

ideal sense, is not actual or potential cause, I. i. 8. It
is but a formula of words or thought until the Forces
themselves are prepared and correlated to act and interact on each other and are put into actuation.

38. Law-Force is the Positive Forces actualizing the
Ideas — this *lifeless* Law. In the highest ideation it
is force, forces impregnate from the Powers of God, and
actualized from the Divine Ideas, impressing upon it,
concreting in it from the fundamental forces, specific adjusting correlations. It is, in each, the transcendental idea
as it is *concreted* in forces. It is the synthetic creating
and continuous energizing in virtue of the idea and the
forces inwoven in the very creating. As ideas are
diversely subjective, in this sense, so in their actualization
they become objective and diverse; they are then the
acts of Deity gone forth from Him, and stand over in an
objective correlation to him, and throughout must maintain, in virtue of their unitary origin, their correlations to
each other and to him. § 10. The Law, as it is called,
always remains subjective as ideas, but when concreted
in acts of creation they pass over, must pass into objective
immanence in the symbols created, or they are only ideas
in their subjectivity in the Divine Self. Passing over
into an immanent objectivity, these ideas concreted in
forces become Law-Forces, (ectype and type,) and rule
the differentiations, phenomenal order, and interplexed
series and articulations of nature and life. Thus the
implexion of forces is seen in the rock lying at rest on
the ground; though at rest, the attractive force is in it
and also in the ground; raise it aloft and take away the
sustaining force, and it falls back by virtue of the *reciprocating* attractions concreted in it and in the earth. And
there are the forces which stabilitated the rock itself, and
the conscious forces concerned in the determinate uplifting.

Here different implexions of forces are at work in these two bodies, and none can say where the creative forces, in their descent into nature, ceased to be moral forces, or how these forces in nature and life are physical, psychical, and spiritual ; but they start in these intelligible differentiations as moral forces from Deity, and they emerge into intelligent and intelligible moral forces in man. These phenomena in nature and life are but in virtue of positive forces, and their abstract laws are but the *a priori*, the transcendental ideas on which Deity implexed his forces and arranged the proleptic immanence and ongoing of his creation, — or nature and life is a materialism or a pantheism. These law-ideas are attained by the processes by which the Self, in its triplicate nature, ascends, transcendentalizes, and gains the ideation of God and his intellective and efficient system of the Universe. *Passim.*

39. Thus the harmonies and the conflicts of the forces in the physical nature, and still more in the conjunctures of the physical, the psychical, and the spiritual or moral natures, become apparent. In physical nature the forces effectuate themselves in sequences and certainties of causes and effects in virtue of the intelligential functionalizations of the forces themselves, in their differentiations, moving and actuating nature. In the psychical life the differentiate orgasmic forces act in a manner akin, if not wholly as cause and effect in nature, as may be seen in the internuncial reciprocation between the afferent and the efferent systems, in the impulsions of the appetites, the ferocity, the constructiveness, and the cunning, &c., when not consciously restrained and brought into subordination, and in mania, monomania, reveries, and spontaneities. But the autopsic powers come in and manifest their discreteness in conscious determinative regulation and control of all these subordinate and subsidiary forces

when so properly used, regulated, and not abused. Now man needs not a command to execute his natural forces; they execute themselves, and, in no metaphorical sense, they grow upon what they feed, and in their exertion and unrestrained growth they keep man as an animal and as a man, and make him more so. But to attain the loftier heights of his spiritual destiny from the depths of his tribal conditions, Command from a superior, submission from the lower, and obedience from the progressively advancing natures is essential to the movement; and when the command incorporates the exact intelligence of the causative forces at work in the physical and psychical natures and the actual and the progressively possible capacities of the Spiritual nature to regulate the others, and when this regulation is necessary to the progress involved, the Command is seen as coincident with the penal consequences inwoven in the natural action of the forces as constituted into the animalistic and human natures of man and producing punitive effects, and as the love which is in the command and working in these disciplinary agencies is evolved, the Command will be seen to be wise, just, and merciful. If the indulgence of lust, cupidity, ambition, pride, and the dark catalogue of human vices and sins, bring direful consequences to the human Self in its system and in society, surely the command is wise in pointing them out beforehand, in preventing them by penalties denounced, in lifting up the Self above the conditions which they inflict here, and which, in the upper light to which they lead, show the stain they impart to the spiritual nature.

Those who can enter into the interior life and, with an observant eye, behold the movements of nature and life, and see the deep chasms and the gloomy intervening distances, with their horrors of vice and crime in battle-

field and brothel, and drunkard's den and felon's cell, and
gilded resorts of æsthetic voluptuousness, in official frauds
and official irresponsibilities, in history and in the ever
recurring present, which separate those who normalate
their lives in obedience to the Commands from those
who will not learn, even by the penal consequences inwoven in the causes and effects of passionate and appe-
tive indulgences, will thank God for his merciful and
loving ministrations in a system resting on Commands,
radiant with the light of that intelligence which comprehends all causes and effects, and a love leading the
children of men to a higher love, yet serenely terrible in
the calm order which exacts that the spiritual man shall
maintain his supremacy over the animal and the man.

40. It cannot but be that the Self, from its position in
the web of its animalistic and human complexures, its
early instruction through its sensuous nature, and its
intercourse with a corrupt world, will be educated by
their vulgar, selfish, and destructive Imaginates, I. v. 26,
and their many vile or voluptuous or degrading habitudes
of soul, in gluttony, drinking, sensuous habits, fashion, and
guilty fame that seeks notoriety, not for the good which
may be done, but for the pride or the gain that may be
enjoyed.   And all these can be conquered only by sacri-
ficing them.   This is the price of victory.   'However
slightly *you* may reflect, you are reminded that the past,
the present, and the future comprise all, and that this
all is nothing.   The past is already past, the present is
fugitive, and the future is not — *and yet it is!*   The
necessitous are overwhelmed with privations, the rich are
satiated with abundance, or wretched in the insatiate
avarice which consumes their lives, the powerful are
tortured with their pride, the idle suffer weariness and
the fierce solicitations of their unoccupied passions and

appetites, the inferior are envious, the rich and the great
are disdainful. The conquerors who overwhelm nations
are themselves overcome by their passions, and they
trample upon others in order to fly from themselves ; and
the multitudes who rush to war but adopt or possess the
passions which ruin countries, and degrade and dissolve
society. Luxury consumes with its shameless ardors the
life of the youth, who, when he becomes a man, is inspired
by ambition and devoured by the flames of this passion.
When luxury and ambition are weary of their victim,
avarice takes possession and gives an artificial life which
may be called wakefulness. Avaricious old men only
live because they do not sleep ; their life is simply watchfulness. Regard the earth throughout its length and
breadth, and consider all that surrounds you, *annihilate
space and time*, and you will find among the abodes of
men only what you here behold — a grief without inter-
mission, and a lamentation that never ceases. But this
grief freely accepted is the measure of all spiritual greatness ; for there can be no greatness without sacrifices,
and sacrifice is only grief voluntarily accepted.' — Cortes.
But how conquer, how make the sacrifice? Only by
the possession of Ideas in all their grandeur and their
universal moral correlations. By early training in the
impressment of beautiful and lovely forms of spiritual
Power, in solemn symbols of the severely serene Truth,
and in the exalted sacrifices of Love, all — all — instinct
and intelligible with the Living Spirit within, and thus
rising up above the vicious and sinful Imaginates, the Self
aspires to the Divine Ideas — to the deific Fore-plan in
the solemn majesty of the Almighty Presence, inweaving
his Power, his Wisdom, and his Love into the Prolepsis
of the solidaric movement of Humanity, wherein His love
in the beginning can only reciprocate to our love in the

End. The Causative Beginning and the Causative End. I am the Alpha and the Omega. God knows; Man must learn.

Idea is therefore properly used in two significations only. *a.* When speaking of or distinctly referring to the Platonic Philosophy, it means "an eternal pattern which subsists according to sameness, unproduced and not subject to decay; receiving nothing into itself from elsewhere, and in itself never entering into any other nature, but invisible and imperceptible by the Senses, and to be apprehended only by the pure intellect." — *Timæus*, ch. xxxi. vii.; § 7, *ante.*

b. Idea as herein used is the complete or approximative ideation of the transcendental system, from before the Beginning, in the divine Omniscience of the cosmical movement wherein the diving creative Ideas were selected from his omniscience and inwoven from his coordinate forces into the appropriate kinds or differentiations of matter, forms of organizations, forces of action, functions of powers and their correlations, — as in starsystems, suns, and their planetary and cometary systems, continents, mountains, coast-lines, rivers, &c., and in the vegetal and animal orders in species, families, and classes on this earth, and men in their diversities with their moral correlations to each other and to the whole, outlining the tracts of migration and history, for the roughhewn destinies of mankind, in which nations and individuals shall write the records of their existence, yet in the consciousness of their responsibilities, in these adjustments, leaving room for human motives and intentions, but in the great movement of the whole the parts are so adjusted, for the intervention of the philosophic and autopsic contingencies of action, that the individual intentions and motives, and the rise and the fall of nations, are controlled in the operations of the general design to the Final Cause. The obtaining of the Ideas, in their syntax of forces for their realization, is the foreplan of a creation and history in outline, in which individuals or nations, acting regardful or regardless of their conscious position, may succeed or fail as individuals or as nations, and yet the ultimate end, inwoven from the beginning in and for the end, shall be attained.

c. Idea should not be used subjectively. Ideates, as a subjective term, will be free from misapprehension. It is recommended by its cognate derivation from Donne's verb to ideate, to form in idea, thus having already its subjective impress, and is in harmony of language, yet is the direct antagonist of all the foul and unworthy Imaginates.

41. Motive, in Intellectual Freedom, is the means or ends, or both, presented to the Intellectivity for its choice and election as a subject and object of contemplation, or to be by it delivered over for execution and accomplishment, and which involve no sense of obligation. Motive, in Moral Freedom, is the means or ends, or both, prompted by some sense of animalistic gratification or human purpose, and which is presented to the intellectivity with a moral alternative in some love, higher up or lower down, for contemplative use, or to be delivered over for actuation in some or any of the deeds of life, and thus always involves the moral obligation to do or not to do for some gratification involved as alternatives.

42. Motive is to intention somewhat as cause is to effect. There can be no intention without motive. This again gives the intellective appetency — the *intellectus appetitus* of Tongeorgi. But as motive is always presented in alternatives, it is not as cause and effect. Intention is therefore the motive *selected* on which the Self

intends and pursues its course of Thought and Action. That there are organizations so constructed, in the human diversifications, which cannot but yield to certain temptations or impulsions, it does not follow that there are not, in them and in others, in the different degrees of their ascent or conditions, points along their lines of movement where Election is their province as it will prove their duty.

43. Sentiment is a passion or affection, or both, moulded by the intellectivity into a more or less consistent and persistent state of feelings, and as such is a motive and influence to conduct, — e. g. a sentiment of honor, of pity, piety, &c.

44. In I. i. 23–25, 35 ; iii. 29, and throughout, it has been shown that the operation of the pure intellectivity can only be concerned in the pure forms of quantities ; in I. 35, 36, 37, and throughout, the Proleptic Morality is brought into view as those moral correlations which are adapted to men in their individual separateness and in their communal solidarity, by which, in this particular planetary existence, progress through the prolepsis up to the divine life alone is attainable ; — in this movement, character, at each step, results from the combinations of the elements brought into action, and it is more or less animalistic, and more or less human, and more or less spiritual, as the one or the other predominates ; and the motive, intention, and sentiment are colored and stained, or translucent as the Light and the Love of God. As the Self passes from the submissiveness to command to the self-normalation of life in the ideational cognition of Deity in his correlations to humanity, its consciously adopted and governing rule and law of conduct is Principle. Principle, therefore, is that growth and modification of those powers of and around the Self, by which it gains and possesses and conserves its proper spiritual life. The

dynamics, integrated as law-forces into the astronomic
bodies, produce the uniformity of causes and effect, as
other differentiations of the forces produce their correlate
effects in their respective physical orders ; but the uniformity of cause and effect does not in like manner prevail in the conduct of personalities in human life. In
animal life their orgasmic forces collect, from time to
time as they are exhausted by use, and appetize when
the occasion and the object occur which solicit or tempt
them, and they are aroused into action and thus act as
cause and effect ; so too in the human appetencies and
passions ; but here the conflict with these orgasmic forces
and the independent action of the autopsic Self in its
more or less capacity of acting from itself, or in its subjugation to the animal, make the results uncertain and
contingent. As man ascends above these conditions, and
.brings them under the control of his Proper Self in the
normalation of Principle, his conduct again becomes stable
and fixed, and therefore more likened to cause and effect,
until it is appreciable that the conduct of a man of
thoroughly inwoven principle has all the certainty · of
cause and effect ; but it is conscious self-directive cause
and effect. It is from this point he can ascend and see
that an Almighty causation, Omniscient in its purposes,
and its purpose one which shall harmonize all his multiplied causations in nature and life, always acts with the
certainty of Cause and Effect. This is Order ; and
moulding humanity by the movement of his great prolepsis to this Order is his Justice, and this order and justice
in the Causative-Love inwoven in the Beginning, and
unfolding and reciprocating in the end, is the Conciliation
— καταλλαγην — of man and God. I. iii. 27–32. The
abstract Morality, as a dry intellective system, as a rationalism, I. iii. 31, without the distinct recognition of God

as a Personality requiring obedience, and without obedience rendered from Love and Reverence, is only a *rule* of action, and its only obligation ! — inducement to adopt it, is Prudence. . Principle is the system of Moral Life adopted for conduct through reverence for the God who has created and correlated the Self to Moral Life as the essence of his nature, and for the love of duty as love of order, in a love to God.

a. In a simple grasping of the Moral Correlations of life by the intellectivity, without any sanctions, given in commands or deducible from the consequences of conduct, the Self can only weigh this abstract and unsanctioned morality against motives, sentiments, and passions of the animalistic and human natures, and act accordingly. When it finds its spiritual nature, it at once must find God in these processes, and the whole force of the question and subject changes to Principle.

b. If to this abstract *a posteriori* morality, I. iv. 3, 4, is added a temporal self-vindicatory energy, in causal consequences inwoven in and injurious to or affecting the organization of the Self, or even if they extend into a future state in one form or another, it forms, also, only a balancing of prudential considerations in which pains and penalties are weighed against wants, desires, passions, appetites, &c., in their burning prurience. Yet the fear of God is the beginning of wisdom. II. v.

c. So far, this vindicatory Reason, only made vindicatory in these causal effects, imposes on this Self the persuasiveness of the ought as a question of prudence. It does give the Sense of Obligation as a Moral Duty. It does not give the obligation of duty. Its only obligation is the fear of punishment, without the Sense of Love which gives the Love and the Moral Freedom of doing the Duty in a love of duty for the love of God.

45. And so Morality, claimed by Rationalism as an eternal system of truth existing in nature and without God, — as the Eternal Principles of Nature, the Eternal, the Perfect, the Absolute Reason, — is but a rule of conduct to be tossed to and fro on the balancing adjustments between prudential considerations and human passions and fanaticisms. And Morality as logically deduced in any system of pure Rationalism with or without the admission of the being of God, omits that counterbalancing element in all human conduct in which the deep and hidden love in men, commencing in orgasmic appetencies and controlled by authoritative and disciplinary sanctions and punitive retributions in individual and national life, unfolds into Principle — a Love of Purity and Holiness. The true and full ideation of the human or proleptic morality therefore involves and includes a power of actuation for good or guilty deeds — an intellectivity to elect between sinfulness and goodness, and to select and arrange means and modes of actuation, and to determine time and place, and a love oscillating between the gratification in goodness or vice and crime, and thus capable of degradation or elevation.

46. Come now with the depurated and full preparation of the Spirit, for where we enter is the Holy of Holies. God is Power. His first manifestations to human Intelligences must be from and by Power in some form of objectiv-faciency setting over nature from God, whereby the Self occupying its conscious, self-objective position can cognize the symbols and forces of this manifestation of Powers. When Powers create Symbols and differentiate forces, and thus give them Forms and Qualities, the affirmative cognition must come that Power is wise — is Intellective. And when this Self perceives and knows that in all sane action of a conscious

intellective and actuative power there is Intention, an
end in some gratification, it must find this in God — as
Love. Power, in some form of manifestation, stabilitation and action, although the last act, as it were, of the
Divine Self concreted in symbol, is the *first* presented
to the human mind. The symbol must be seen — in
the senses — before there can be any inquiry or cognitive
perception of the wisdom, the intellectivity by which it
was fashioned. I. v. 6–15. And in like manner the
correlations of the thing and its adaptations to use and
gratification must be intellectively and intelligently obtained before the Love in the uses and ends can be appreciated. The Self, in its threefold powers, thus intuscepts the divine forces in their Trinity and in their
ontologic coördinations as Unity. I. v. 39. Thus God
presents himself in the intelligible symbols and forces
of nature and life to the intelligent self on this side of
nature, and in this presentation he is found, in all his
acts, a triplicity of Powers. All of his coördinates are
Powers, and none of them are empty and lifeless abstractions of the Rationalisms and current theologies.
Thus we ascend and find that God is Creative Power,
is Creative Intellectivity and Creative Love. Wherefore
should Powers create? wherefore should Intellectivity
count and plan and arrange the infinitesimals in their law-
forces for and in the protozoa, and in the complicate arrangements of all life and nature, and hang the stupendous systems of the illimitable worlds on the action of
his Forces? Infinite, Holy Love — or else there is no
God! I. iv.; v. 31–41. Dr. Nathaniel Emmons, a
distinguished New-England divine, says, (Park on Atonement, p. 116,) " All the moral perfections of Deity are
comprised in the pure love of benevolence. God is
Love. Before the foundation of the world there was no

ground for considering the love as *divided* into various
and distinct attributes. But after the creation new relations " (correlations) " arose." Before the foundation
of the world there was no *object* for this love in it, and so
far it was not benevolence, and Love, therefore, subsisted
only in its coördination with its coessentialities ; and after
the " foundation " only it was divided, and so could be only
in virtue of these coördinations ; and distinct attributes
are only the subjective ideas in the Self for these established correlations thus effluent and objectified out of
these coördinate Powers. But this language, as that of
all sound philosophy and theology, implicitly affirms the
simple ontology of the Love. · I. i. 20. Dr. Maxcy, *id.*
94, says, " The Law, whose essence is Love, tends in its
nature to secure the highest happiness to all rational
creatures." This is the universal expression of all advanced thinkers. This Love in the Law, *ante,* §§ 35–37,
45, is adjusted to the correlate love in the rational and
loving creature for whose discipline and education it was
instituted. Deity is thus presented in his ontologic co-
essence as Love, — man comes afterward, but the Law
must be seen, as preëxisting in the divine intellectivity
and moving into the order of his great arrangement, or
else it is *ex post facto,* and as such harsh, cruel, and unjust ; and this very view destroys the conception — the
ideation of God as a God of previsory Wisdom and of
provident Love. Love, Wisdom, and Power preceded
the creation of man, and they enter into all his works,
not in the perfection and harmony of his precreative
coördinations, but in the unfolding grandeur and glory
of his appointed prolepsis for the conscious intusception
of his obedient and loving children. So man learns.
Their joint effluence is the spirit which he has poured
out on all his works, (Proverbs viii.,) and the great day

of his prolepsis is manifesting it. This view is necessary to any theology, and no philosophy which accepts a God can, in the light of these Three Essential Powers, otherwise find a root for their origin or construct a system of the universe. So Donoso Cortes, in the processes of philosophic thought and under the sanction of "one of the most eminent theologians of Paris belonging to the glorious school of the Benedictines of Solesmes," says, " Providence is a universal *Grace*, in virtue of which all things are maintained and governed according to the divine *counsel*, as Grace is a special providence by which God takes care of man." All these views are phenomenal, but they all imply underlying ontologies from which Law, Counsel, and Grace are effluent; I. i. 17–23, 18; title-page; and Law and Counsel can only be found in the Divine Intellectivity, and mercy, benevolence, beneficence, Grace in Love, and both only as wise and just in their coördinate union for Actualization, and to be harmoniously justified in their Causative End. In God the one element cannot be found without the other; and again the creation and the government of God is the unity in manifestation of his coördinating Power, Wisdom, and Love, distributed in the Past-Future-Now of his omniscience for the order of his prolepsis throughout the flowing ages. Grace is, therefore, that arrangement of his counselled Love, in the order of this movement, by which the Love in the Beginning reciprocates to the love in the progress and to the causative-end which is in the Love at the End. I. i. 15, 18; iv. 20; v. 24. Its end is the causative-end in God, and it is the end for aspiration in man — his attractive, causative end. See the kinhood of §§ 41, 42, 43, 44.

The instinctive and the determinate language of life, philosophy, and of religion, as to the causations and offices

of the Attractive Love, may be gathered from the following authoritative remarks. " The Holy Ghost, in conveying to us an idea of perfect love in the Scriptures, generally employs terms expressive of *union*. Thus St.
Luke says, the multitude of believers had but *one heart
and soul*. Acts iv. 32. Our divine Lord prayed for
the faithful that they may be One. John xvii. 11. St.
Paul desires that we should be careful to keep the *unity*
in the *bond of peace*. Eph. iv. 3. This union of heart
and mind is certainly a mark of *perfect love*, since it
unites several souls in one. It is said ·in the Holy
Scriptures that the soul of Jonathan was *knit* with the
soul of David, and Jonathan loved him as his own soul.
The great apostle of France, in describing the properties
of Love and quoting the opinions of his master Hierote,
repeats above a hundred times in one chapter that Love
is unitive, that it *unites, assembles, collects, and com-
presses* all things, reducing them to unity." — De Sales's
Love of God, B. I. ch. x.; I. i. 17–23; v. 31, 33, 34.

47. In the organization of man brought into view it
is seen that he is a Cognitive Spirit in a complexus of
somatic and psychic organisms moving into action, and
eliminating his self-conscious Powers and mastering his
orgasmic forces; that his only *means* of•cognition are
his various senses, but that he ascends from these to his
intuitions and to his intusceptive · ideations; that from
his intuition he only gets · the Insistent Truth, and from
this, in his *pure* intellectivity, he moulds the Formal
Logic; I. i. 34; iii. 29; that by his intusceptively ideative processes he catches the Proleptic Morality, and
from these he moulds and lives the Moral Logic. I. iii.
29. He thus gains the prolepsis for the creation of the
world, the successions of the geologic eras, and the tribal
and historical movement of the races of men. Thus

man attains the Suprasensible. He only can so attain
it as he lives it and inworks it into the very roots of his
existence by the accumulated experiences of the ages;
or as he may be creatively formed for attaining it in
shorter time and with less disciplines of tribulations.
I. v. 38. Any Prophecy, Revelation, and Inspiration is
only possible through these means. The distinctive
mark between God and Man in this aspect is, God
knows and Man learns. It will be found the surest distinction between Creation and Emanation or Pantheism.
It is the key to the mission and the method of Christ.
Prophecy is impossible without a prolepsis, more or less
general or special, in which the coming fact or man, fore-
told, is prearranged; — and this prearrangement is intimated in the location of oceans, continents, islands, in the
construction of mountain ranges, rivers, and coast-lines.
§§ 40, 39, 35–38. Revelation is only possible or appropriately probable in reference to some fact or facts in this
prolepsis. And Inspiration is only possible or appropriately
probable in reference to some truth or doctrine connected
and inwoven with the facts and pertaining to the conduct
and moral destiny of man as the subject-object of the movement, and which is given above the method by which
man, in his natural order, gains his ideative cognitions.
Revelation and Inspiration are possible only in the higher
organization or higher exaltation of the organisms, produced by the coöperating agencies at work in history
preparing the agent receiving the Revelation or the Inspiration. Preparation in some form is essential to him
who receives and transmits, as *mentalization* is necessary
for those to whom the communication is to be made and for
whose use the communication is given. Or else, a constant
miracle is evoked to account for the prophecy and the
revelation and the preparation of the races to receive it.

Such a constant effluence of the divine powers would be God permanent in Humanity, and confound all just conception of moral cause and effect. I. i. 4, 17–23 ; iii. 30, 31. This would be pantheism, and destroy the moral order of the universe. The higher ideation will behold what the lower cannot, and greater capacity for ideation will intuscept and comprehend what its capacity for ideating can reach, and not anything more. I. ii. 17 ;. v. 20. If revelation or inspiration gives any more than this, they can only give it by increasing the capacity of the cognizing Self, both of that one which receives and communicates and that which receives. Each must perceive and know for itself, and this only as higher or lower powers are vouchsafed, or as it co-works with the order of the Almighty, and normalates its life to the upper light, and catches the Light and Love of that Providence which is a universal Grace, and which accords to the conscious co-worker with him the power of further progress in the conscious causality of the sympathies for a diviner life ; — "It seemed good to the Holy Ghost and to us." This defines Revelation and Inspiration, as it gives the human means and modes, again in turn, for analyzing, synthesizing, and giving rational form, and imposes the responsibility for making the progress which they imply and unfold. Thus they are inwoven into the life by the conjoint action of the triplicate powers in the Self, concurring to the divine order, and giving the purification and the perfection of the theologic Will. I. i. 17, 18. This is the Mentalization of the Nations.

48. The crowning height of four diverse divisional movements of the Supreme Mind, with many beginning orders of differentiation in the geologic eras, in animal creations, is Man. He is the conscious autopsic Ego,

the King and the Slave of all. The elements are all mixed in him ; the dust upon which he treads and feeds, the rock which he calcines and casts on his fields that he may live from their fertilization, the grass of the fields, the grains, the fruits, and the very animals prepared for his use by their autonomic growth and their instinctive adaptations for his use, the air, the water, and the light by which he and they grow, and the animal appetites, wants, passions, and instincts, with wants, passions, and fears and loves and hopes peculiar to himself, are all inwoven into and around his nature ; and central in all this complexity stands this Self, this conscious Ego, and out of this maze by his own conscious self-normala- tion, responsive and aspiring to the Love which over- souls all, he, each, must rise on the force and dignity of his spiritual nature, by obedience to the command en- forcing purity and holiness, and the normalative inwork- ing of that Charity which gives a new purpose to his unfolding life, — or he must remain forever fallen, wrapped in the foul cerements of his animalistic and human natures, responsive to and increasing his moral depravity by the deadly infusion of his conscious forces into these orgasms, augmenting their intensity.

This gives the Self in its entire scope of exploration and of labor, of Spirit and of Soul and of Body. He unites all the characteristics of King and Slave ; toiling through the mazes of the great labyrinth, there is much that he must know, but in pain and sorrow he must gather his dearest, noblest knowledge ; he must rule his own passions, and be ruled by the rightful and inexorable command, yet given in love ; he must ascend to the clear cognition of the Divine Ideas, and see and live them as they become Law, and conform his own life to their system ; he must govern his own love of wrongful

gratifications, or be subdued by them ; he must enthrone
his own reason in a depurated love amid the unyielding
steadfastnesses of the eternal Principle and the appointed '
moralities, or it shall be dethroned by the busy cunning,
and restless intrigues of fraud, chicane, and expediency ;
to receive mercy, he must show kindness, yet in the
order of the Almighty which closes around him at every
step ; to be forgiven, he must forgive ; to receive of the
charity which may cover the sins of his threefold nature,
he must exercise that threefold nature in consentaneous
deeds of Charity ; from the Primal Source he must receive and inweave Love into his own heart, to be worthy
of the Love which giveth so much ; to attain moral
elevation he must battle like the humblest in the great
empire of man, where God governs against a host of
vices, follies, and infirmities ; standing on the topmost
summit of the animal‚generations, and enthroned in the
conscious magnitude of his willed force, of his intellectivity and love, he still sees Powers and Dominion above
him, which he, in humility and justice, gentleness, meekness, and love must confidingly serve, in the discharge
of his proleptic duties.

This ‚ King-Slave, the Self, pervading all the conscious activities in loving and intellectualizing and in the
deeds of life, is the immanent solidaric agent, identifying
itself in its continuous and articulated cognitions, and
which must scrutinize the facts and declare their philosophy, and in obedience and hope be lead by command
and law to the Freedom of Love.

This conscious Ego is the immanent Self in its own
solidaric activities. With ability to actualize in positive
force, with intellectivity to declare determinately its
mode and means and end of actuation, and to control its
self-forces to its own elective ends, and with loves in

20

diversity and complexities of gratifications to be consciously gratified, or withheld from gratification as the ends of its thought and action, it stands amid the operations of nature, and touching on all sides the filamentary correlations which unite it to other selves and all to the Unitary Source of origin, it is within the limits of its own allowed circle, a lover, a thinker, an actor — maintaining good or projecting its own created evil into the series of moral causes and affecting them to the third and fourth generations, and until in the deepening currents of the vicious multitudes the vindication of the Law and the Command is written in the blood and desolation of nations.

BOOK FIRST.

MASTERS AND WORKMEN, AND THEIR WAGES.

——◆——

CHAPTER SEVENTH.

LAW, PROGRESS, VICE, EVIL, SIN, JUSTICE, GOD.

1. In I. i. 35, 36, 37, the Insistent Truth, the Divine Ideas, and the Proleptic Morality are brought into view, and, throughout, the validity of their intuition and ideation has been demonstrated as the Forms by which God created all things, and as giving the correlations which bind the whole in their system one to another, and to himself; in vi. 40, 41, 42, 43, 44, Ideas and ideates in their necessity and influence in the formation of motives, intentions, principle, and for the elimination of the proleptic morality as the rule and obligation of conduct, are placed in their appropriate system, and as they present the alternatives for the Intellectual and the Moral Freedom, vi. 33, 34; in vi. 35, 36, commands in their subjective and objective significance are defined; in i. 41, Transcendentalism, as the Standpoint above and prior to the facts, the deeds, the actualization of creating is obtained as the foreplan of the creation; in i. 42, Prolepsis, the divine transcendental movement by which the creating is objectified forth from Deity in its system of forces, is given in general

terms; in i. 34, the Solidarity of the Individual and of
the Races is found by that regulative law which gives
cause for effect, and underlying forces or ontologies for
phenomena, iii. 20; in i. 26; ii. 7, 8, 14; iii. 3–5, the
Autonomies of the individuals, and of species and races,
are presented in their underlying formative forces; in i.
31; ii. 17; iii. 9–16, the conscious Autopsic Self stands
forth in its independency; in i. 34, the Communal Solidarity of the Race is determined; in i. 33; iii. 9, 14,
Spontaneity and Normalation are described and contrasted
and set forth as the movement-forces of individual and
historic life, and these spontaneities as being controlled
by the Intellective force; and in vi. 47, Grace presides
over the order and harmonies of the world, and is that
• everlasting counsel of Love whose light guides, whose
power sustains, whose love nourishes, and whose gift, in
the fulness of these powers, is immortal youth; in v.
47, Prophecy and Revelation appear as that exaltation
of the ideative process by which facts arranged in the
prolepsis yet to come are foreseen, however this exaltation
may be derived or given, and in like manner Inspiration
can only be of doctrines and moral correlations, then or
yet unknown; in iv. 24, and in the whole of the processes
evolved, it is seen that the New Birth into the Spiritual
Life is the normalation of the whole life of the Self in its
entire complexure of body, soul, and spirit, to the Knowledge and love of Holiness, and the Spiritual presidency
of the Self; now Sin must have reference to all these
subjectivities and objectivities, and in the system of life
as, throughout, is set forth.

2. From these grounds or data, the idea of Law, not
only as a rule, but as a command to obedience whenever and however they are clearly presented to and
obtained by the Self, must be raised and perfected to the

consciousness of the Self, and then Sin will appear as a conscious, and vice as an unconscious breach of this Law. The following definitions of Law have been submitted by various jurists and men of ability : —

a. " Law, in its most comprehensive sense, signifies a rule of action;" and is applied indiscriminately to all kinds of actions, whether animate or inanimate, rational or irrational. This, then, is the general, verbal definition of law. Such a definition can give no philosophy or science. I. vi. 37, 38. In attempting to build either by the use of this term, it means something that is the cause of effects in different senses.

b. " Law is a rule of action dictated by some superior being."

c. " Municipal Law is a rule of civil conduct prescribed by the supreme power in a state commanding what is right, and prohibiting what is wrong." — Black. *Com.*, I. 38, 39, 44.

d. " Lex est ratio summa, quæ jubet quæ sunt utilia et necessaria et contraria prohibit." — Coke, I. 17. Law is the Supreme Reason, which commands those things which are useful, and prohibits the contrary.

e. " Lex est justorum, injustorum distinctio, quiddam eternum in mente Dei existens ; recta ratio summi Jovis." — Cicero, *De Leg.* lib. I. et II. Law is the distinction between the just and the unjust, existing as an eternal insistence in the Divine Mind. It is the right reason of Supreme Love.

f. " Lex est regula actuum moralium obligans ad id rectum est." — Grotius, lib. I. c. 1. Law is the rule of moral actions obligating to that which is right.

g. " Law is a rule which an intelligent being setteth down for the framing of actions by." — Hooker, *Ecc. Pol.*, B. I.

h. " Lex est sanctio justa jubens honesta, et prohibens contraria." — Braxton.   Law is the just sanction commanding the honest, and prohibiting the contrary.

i. " Law in reference to moral actions expresses the sense of the law-giver as to what is right and the value of the right." — Barnes, *Atonement*, 80.

3. Law has been brought into notice as the *a priori* — the antetypal ideas, and their arrangement in their system of correlations, by which things and forces are to actuate when the law-forces are inwoven into them and put into operation. I. iv. 3 ; vi. 37. It has also been noticed as the deduction which is made *a posteriori*, as a mere generalization from the facts and forces in existence. Neither of these in their simple apprehension can give the ideation of law to rule the conduct of the Self. They must both be carried further into consequences as flowing from forces which may or will enforce their observance in their effects. If this view goes no further than the perception of forces executing themselves, then there is no moral obligation to observe them other than to mitigate or avoid these evils, or to choose the alternative of gratification, and the evil effects of the gratification which the exercise of these forces as passions or indulgences will or may produce. I. vi. 44. To give them the validity and the obligation of Law in the sense in which this term should be always used in reference to conscious autopsic natures, Law must be seen as these preëxisting and antetypal ideas arranged and correlated to penal effects for the government and welfare of those creatures to whom they are thus adapted in the creations which are produced and thus made the subjects of these ideas as laws. This is the Common Law of man as a responsible agent. It may be likened to that great body of Common Sense which men must exercise in society and government

where there is no positive law to regulate their conduct, and which courts of justice in their wisdom enforce in the absence of positive or statute law. Yet in this view Law is seen from the first, in its antetypal institution, as a positive command, yet as something which he must obey other than for the penal consequences connected with its infringement. He must obey it for its effects on his own welfare, and on the members of society with whom he is associated and thus correlated in giving moral life, and which he should obey for the Wisdom and Love it implies and demonstrates. Adopting Law then in this signification, Law as it is applied to an imperfect moral creature is not only a means for disciplining moral vice and punishing crime, but it is a process for preventing by educating to a knowledge of the vice and the sin in the individual and the race which the law implies, and for unfolding a love of the order, justice, and righteousness which, in the perfection of life, is its unfolded mentalization. And this order, this justice, this righteousness, cannot coexist with the moral vice and sin which is thus the subject of the implied or the express inhibition. The definitions set forth are therefore imperfect, and do not give the elements from which the nature of Sin can be collected, for —

a. There are sins which are not defined or even suggested by the terms "rule of action," for the animalistic nature of man has one rule of action, in the sense that this phrase has always been interpreted, a law-force executing itself, and the moral nature of man another rule which he must wisely and lovingly adopt, not only for his *rule* but his love of action. The first, of itself, acts as unconscious cause and effect, and the latter as self-conscious cause and effect. I. vi. 32, 44. The sinfulness of the former may depend on circumstances, for

adultery is prohibited both by this common law of nature
and the wise command which conserves and builds up
the moral nature of man and the system of society. I. v.
33. In the former, in the vegetal Kingdom, the continuance of the species is by the inherent law-forces
executing themselves, upon contingencies, yet which are
so closely inwoven in the correlations of pistil and stamen
that there is scarcely any chance of failure ; in the lower
animal orders they internuncially execute their offices ;
and in the higher they always tend to execute themselves,
yet are not self-conscious, and in man they are inwoven
into his organization, so that they are in constant inosculation with this self-conscious autopsy, and are to be ruled
by him into moral order. The law of the one is the law-
force which executes itself; in the other it is conscious
self-direction which executes as animal, man, or self-
normalative Spirit; and the same term applied to both
is a confusion of thought and a want of analysis.

b. In b, the same confusion prevails, for both conditions
are equally prescribed by a superior for the rule of action
of the inferiors, and no discrimination can be made without higher distinctions.

c. Nor does the Municipal law always command simply
the thing which is right, and prohibit that which is wrong
in and of itself; nor does it pretend to this, for its laws
are founded not so much on that which is *malum in se* —
wrong in and of itself, as on those intentions and sentiments of artificial policy which require men to sacrifice
their highest convictions of what is right, both according
to this Common Law of Morality inwoven in the order
of the universe, and in the Commands which tend to its
higher cognition and its more authoritative sanction and
obedience. This artificial policy is constantly changing
with the changeful forms of government, and with the

policy and ambition of artful and unscrupulous men, governed by human passions and maligned sentiments. The law regulating property is always an artificial system, depending more or less upon the policy of the government, and this more or less upon the condition and progress of society. Its extreme artificialness is exemplified in the fact and in the many changes of the Feudal Law in England, and in the effects and influences which it has upon all questions of *rights* and remedies in England and America at this time, and in the diverse forms it has assumed in the various governments of Europe, regulating successions and remedies, &c., &c. It is the result of artificial systems, which recognize and enforce these artificial Rights of individuals, and permit one man to recover one sum for a day's labor, and another a different and a larger sum, and without regard to the moral condition or wants of the parties. It is strictly artificial. It enforces the gambling contracts made on the fluctuations of prices, whether of gold or grain or other things ; and if the monopolist can foresee a scarcity of the necessaries of life, and can gain the control of the market, this law not only protects his property thus obtained, but enforces his contracts at exorbitant prices produced by the scarcity which he himself has occasioned. It protects the title to land which was obtained a few years ago by robbery or fraud from the Indian, and secures another title by the statute of limitations, and which was commenced in personal fraud, or in an adverse possession against the legal owner. Everywhere it is an artificial system, directly the reverse of any abstract horizontal level of equality or theoretical justice. And the whole body of the Municipal Law gives no definition of *Justice.* Yet Law is necessary to the wellbeing and moral and intellectual growth of the Individual and of Society ; but it must be, in the conduct of individuals and the movements of governments, subordinated to the Proleptic Order and Morality of the Almighty, or vindicated in revolutions and reconstructions. Deal wisely with the paradox.

d. Nor is it simply in " the highest reason which commands the useful and the necessary, and prohibits the contrary," for this omits or only implies the subjective sense of obligation — the moral principle of obedience, and makes law to depend on the positive power which commands, relegating the subjective Self of the individual as the recipient of the command and as the conscious respondent to its requirements, in his intelligence comprehending the value of the law and loving the law for its intrinsic value. This definition is *logically* correct when applied to Deity as giving law, commands to a full autopsic agent wholly comprehending the law in itself and in its consequences of obedience and disobedience, and with full freedom or equal election for the obedience or the disobedience. It wholly omits that view of Law or of Institutions wherein intelligent and loving power is exercised for the control of unmentalized natures for the suppression of their animalistic and human impulses to action, their undepurated motives, sentiments, and intentions of conduct, and this for the unfolding, the disenvelopment of the spiritual powers into a love of order, justice, righteousness.

e. It is not merely in the distinction of just and unjust which exists eternally in the divine Mind — the *recta ratio summi Jovis* — for this gives no sanctions ; and there are forms and ceremonies and observances in society, in governments, and in all religions, which are not wrong or right in and of themselves, but which are simply disciplinary and educative, and which may be wrong or right as means and instrumentalities for reaching a knowledge

and a life of that which is right in itself, for educating
the individual or the tribe up to the full stature of autopsic manhood. I.'vi. 44 ; II. viii. vi. vii. And these
will vary, at least in application, as the progress in the
fulfilment of the prolepsis varies.

f. and *h.* For the same reason it is something more than
the rule of moral reason obligating to do that which is
the Right in and of itself, — as the Ceremonial Laws of
the Hebrews were of temporary enforcement, many of
which were applicable to a small territory and a limited
population, leading that people to the recognition of the
permanent obligations contained in the Wisdom and Love
of their Decalogue. I. vi. 47, 46, 40. And so in domestic
and social life, and in all paternal governments where the
moral destiny of the people by the repression of vice and
the promotion of intellectual and moral freedom is an end
proposed, these, in some forms, must appear. So in the
culture and progress of rude tribes.

g. This assumes or omits the essential morality and
loving care, I. vi. 36, 37, 44, of the Intelligent Power
" setting down the rule for the framing of actions by ; "
and when the rule is to act objectively on others, it
assumes their intelligence and regard for the rule, and
does not provide for the want of them, nor for a progressive mentalization and movement.

In *i* it confines strictly as to what is the Right in itself and the value of this Right, and omits.wholly the
educational means and processes for attaining to cognitions of the right and the fact of normalating the Self
and of being moulded by discipline, instruction, and
education up to the knowledge and the love of Right ;
e and *g.* The love of the Right is the value of the Right
to the Self. The love of the Right, as the universal condition of the race, is the full and perfect love of the

Right for itself. Where all are in the knowledge and love of the Right, it is order, justice, righteousness ; and the law, as an objective instrumentality, ceases its operation, except. to prevent each Self from a "fall" from this Order. This would be Moral Freedom perfected. To a humanity like the vicious and sinful inhabitants of this planet, the education of the movement is that they may attain their end as Humanity in the knowledge and Love of the Right, and this is ordinated in a prolepsis in which all the elements brought into view are means educative to the end. Hence as God is the creator of man, and he has assigned a proleptic morality for man discernible in the moral correlations of life — this Common Law of Humanity — and in statutory commands giving more definite significance to this Law, the end of man is the attainment of the Moral Life, through the means and processes instituted by Him, for the knowledge, the love, and the practice of the Right as it unfolds in the Movement.

The insufficiency of language to convey the wide and solecistic distinction, whenever the attention is turned to it, which subsists between Law as a rule of conduct which the Self must observe or violate on its own sense of responsibility, and Law as differentiate forces inwoven in the positive causes of nature and life, which, each in and of itself, tends to its own execution as cause producing its own specific effects, is seen in Saint Paul. " Now if I do that which *I will* not, it is *no more* I that do it, but Sin that dwelleth in me. I find, then, a Law that when I have a *will* to do good evil is present with me, for I am delighted with the Law of God according to the inward man ; but I see *another law in my members*, fighting (*repugnantem*) and captivating me in the Law of Sin that is in my members." — Rom. vii. 20-23

Law, *lex*, νομος, is the term throughout his whole discussion, and it is clearly used in two several distinct meanings : namely, in the first, as the rule of conduct of the
inward man by which he is to elect his line of conduct on
his sense and belief of his responsibility ; and in the
other it is the orgasmic forces in his members, which, in
virtue of their active, appetizing, and impelling momenta,
are inciting to their respective gratifications. The one
is the rule of conduct, the other are those orgasmic forces
in the members which, by the rule of conduct, must be
resisted and rejected. One is unconscious cause in life,
yet acting, in virtue of the autonomical organization into
which as orgasmic or psychical forces it is inwoven,
directly on the Consciousness of the Self, the inward man,
I. i. 27–34; and the other is the conscious control by the
autopsic self of the cause — *the forces in the members* thus
differentiately impelling to their specific gratifications.

4. There can be no true ideation of Sin without reference to the antetypal ideas and their pre-creative correlations, and then as seeing these as having a positive institution and a conformity in that organic constitution of man
by and through which the Moral Life of the individual is
to be attained and consummated in a love of obedience,
conforming and reciprocating to the Love which rules
and ordinates the movement. This prescribes and legit-
imates the intermediary processes of discipline, instruction, and education, and gives the sanctification of suffering and sorrow. I. vi. 40, 48. Law is therefore the conforming of the Self, in its triplicate life, to the perfected
order and harmony of its own essential correlations.
These higher ideations come with more or less fulness,
and more or less late in life, according to the relations and
the correlations of the particular Self in its order in the
great movement, and the conscious normalation of the

life upwards toils on slowly after. "The Kingdom of Heaven is like a grain of mustard-seed, which is the smallest of seeds, but when it is grown, it is the greatest of herbs." It has its laws and its forces of growth.

5. The divine coördinations move into actuation only in obedience to, only in harmony with, the pre-creative ideas. I. iv. 23–25 ; vi. 46. These coördinations do not move simultaneously in the human sense of the same time, but follow in successions in human time, for to Omniscience a thousand years are as one day, and one day as a thousand years in its cognition, I. vi. 41, b, 46, and in the world-structure the Powers as objective-facient most forcibly and conspicuously appear, and in the lengthened intervals of dark and gloomy periods, evolving from the geologic successions and the historical deployments, the light of Wisdom — of Intellectivity beams and shines through the sweltering masses of terrific powers, and only in the last ages the Love appears and uprises through the order of things, struggling, as it were, to its final and supreme manifestation. So in the history of the races of humanity, so in the Scriptures, God reveals himself, first, as Elohim, the Almighty Forces, and throughout the earlier manifestations his Powers are the object of awe and submission and his chief manifestation ; in the geologic world these powers cease their more forcible demonstrations, greater harmonies appear in the action of nature and reason, intellectivity dawns on the pathways of nature and of history, — and this continued until, in the concurrence of the ages, the Wisdom, the Logos, in its proleptic time, lightened up with its Intellectivity the Way, the Truth, the Life which man must live, but which, like the apostles of old, he cannot live by *his* intellectivity alone, nor until the Love, in its genial unfolding or in the gradual

accretions from the grief of life, shall give an object, a content, an intention, a motive, a sentiment, and a principle — in an end of attainment to the Spiritual Self. But the coördinations move into action in conformity with the pre-creative system. Obedience, then, to the proleptic Morality is the perfectibility of the earthly moral agent; but this is at the end of all his struggles, and purest and gentlest and self-sacrificing normalations of life. II. viii.; III. xi. This obedience is the result of many subordinate processes as imposed in the actual strifes and discords of individual, domestic, social, political, religious, and moral life, as differently modified by tribal autonomy and geotic causes. Therefore obedience in the subordinate condition, so as to attain the final obedience in the love of Final Truth for the discharge of Duties, is necessary to the consummation of the character of the ascending and unfolding Self. Hence all means in the ascent for obedient purification from the conduct produced by the impulsions of the animalistic orgasms and the mere human and prudential motives of action, (and both continued in their filthy and selfish imaginates,) as the temporary and progressive forms of law and obedience, are to be brought into action, and many of these will be seen as always necessary. It is true of the moral life in its progress as it is said to be of language, " everything that is abstract in language was originally concrete, and languages are formed by a process, not of crystalline accretion, but of germinal development," and most assuredly of normalated perfection. " Every essential part of language existed as completely " (although implicitly) " in the primitive germs as the petals of flowers exist in a bud before the mingled influences of the sun and the air caused them to unfold ; " and the spontaneous and the normalative processes of the

human autopsy unfold the life of the germs — the spiritual
from the concrete, in both the animalistic and human, and
ever and forever in its redactive processes, still recloth-
ing — rehabilitating them in new but more transparent
forms. Yet in language it is seen that this rehabilita-
tion is but the new covering of the new thought and life.
II. i. ii.

6. This is the plan of salvation — of elevation from
the envelopment of the animalistic and the human natures
to the presidency of the clear autopsic Self in the accordance of its spiritual life to the higher life of knowledge and love and duty, which is but the actuation of
both. This can only be attained by man in a spirit of
meek and gentle and long-suffering obedience. Take the
dark catalogue of terms — of root-germs, as they have
welled up from the animalistic and human passions and
appetites, and their diversifications of moods and component significations, and contrast them with those few
simple, pure, chaste, and solemn words assigned to the
use of the spiritual life, and their difference, antagonism,
and eternal war are revealed. Fact and language show
that obedience in self-control and self-adjustment of the
powers in the Self to the ideated Divinity, I. vi. 46, 47,
is the law of human progress and final conciliation. The
two terms, in the full sense of all the subjects, objects,
and processional movements, are convertible. It is the
normalation of our instincts and spontaneities, whereby
the somatic, the instinctive life is subordinated to the
human — the psychic life, and the human to the spiritual
— the zoic life. II. ii.

7. Yet elevation of moral life is only possible in the
fact and exercise of some Sense of moral responsibility.
I. i. 35–37. What is the Sense of Moral Responsibility?
It is a subjective condition of the Self which is different

in the child, the adult, the middle-aged, and tne elderly
man ; it is different in Africa, Asia, Europe, and America ;
it is not precisely alike in any two human creatures ; —
yet in all it is, at *base*, the same, and at the base it is
something struggling to get *free*, and thus to think and
love and act in a higher form of life. It is something
capable of change and of improvement, or capable of unfolding disenvelopment, as it is capable of loss or greater
obscuration in the self-indulgence of the animalistic and
human appetencies. I. i. 33, 34. It consciously enlarges
its own boundaries, and they are enlarged by instruction
and discipline, and in a certain condition of education in
society and in certain trains of thought in the individual
are almost always enlarged by the vicissitudes and the
sorrows of life, — nay, however enlarged the theory of
life, and however firm the moral convictions may be, they
are much other and different after realizing them in
tribulations to what they were before. At base it is
something which requires and acquires knowledge —
cognition of *the* Right, and an unfolding and an unfolded
Love of the Right, and in this Knowledge and this Love
to do — to actuate the right as unfolded in the Self in the
proleptic — the progressive order of the Almighty. The
first consciousness of the Self, in life, is of its envelopment
in its inwoven connections with the somatic and psychic
lives, I. iii. 5–15, where it finds these burning animalistic
and importunate human appetencies seeking and impelling to their respective gratifications. Now it is out
of and through, but not intrinsically from this cluster of
passions and appetencies that the Sense of Responsibility
arises ; therefore, what is it ? It is the Spirit grasping
at *a freedom* from these impulsions and importunate·appe-
tencies in which " we all had our conversation in times
past, fulfilling the desires of the flesh and the mind."

21

As this knowledge and this love of *the* Right, II. vi., increase and unfold into the ideation of a divine purity, the
desires of the flesh — the animalistic life and the human
appetencies lose their controlling importunance, and as
the Self subordinates and, in turn, controls these, so far it
becomes free, and in a point in this line of progress it
becomes conscious of its efforts to become free, and the
settled and constantly progressive habitudes of these
efforts ripen, under the conscious normalation of life,
into Principle. I. vi. 41–44. In a fully unfolded love
of moral purity, the Self, thinking the right and doing
the right, loses this Sense of Responsibility in the very
perfectness of its knowledge and love. Love is the fulfilling of the Law ; Perfect Love casteth out fear. The
• Spirit is placed amidst the organic instincts, passions,
affections, and with organic powers of intellection akin
to the instinctive intellections of animals, as cunning in
the fox, constructiveness in the bee, the wasp, the beaver,
the nest-building and singing of the bird, the teachable
sagacity of the dog, the horse, and the cow, &c., but it
uses — consciously normalates the whole. Without these
instincts, passions, affections, and organic intellections, it
would not be man. And when, by the conscious exercise of its spiritual powers, unfolding its Sense of
Responsibility, it clears up and escapes from these influences and organic powers, it ceases to be what it was
before, *but it does not cease to be.* I. iii. 29. The whole
argument of materialism, and surely of spiritualism,
affirms that it does not cease to be. No substance or
essential Force is ever lost, but on the contrary, in this
instance, the powers of self-government, of autopsic *
normalation, are constantly increasing, and become more
manifest as long as the organisms remain perfect to respond to their conscious action. As the Self ascends to

this supreme power of self-government, it only attains
its higher and clearer life. The worm that crawled
upon the earth now flies and moves in the sunlight of
heaven. As this Self thus acquires powers to act in its
spiritual freedom, it escapes from its animalistic and
human environment, and unfolds until this Sense of
Responsibility is the perfect freedom of its nature to act
according to its exalted or disenveloped spiritual existence. So far and in this way we can follow the Spirit
to its clear Subjective Identity. I. ii. 14–18 ; iii. 25, 17,
20. There it is stripped of all adventitiousness, or, as
the logicians call it, *accidence,* and there the humblest
reasoner who can follow these processes will see it as
Spirit in its triplicate unity, above material organization,
yet using these organisms as its material instrumentalities. And nothing higher or more fundamental than
this can be conceived except the prime creative Forces.
These are the basal elements of all freedom. Man may
beat around the dark walls of his animalistic and human
passions and loves with their endless conflicts within and
the fierce struggles of life without, and he will only escape
from their gloomy and burning environment as he catches
the light — the Knowledge of Moral Freedom in the
Love of a higher life, and taking this Sense of Responsibility so constituted of this knowledge and love, and
holding it as the clue to guide himself, as he follows
it from this labyrinth of passions and appetencies, fanned
and exacerbated, as they are sometimes, to flames of
fanaticisms, he will arrive at length at the calm and
serene life of Moral Freedom. It is only in the system
of a true life of the Self that its Moral Freedom can
be cognized and attained, for any other supposed freedom is but the degradation of animalistic passions and
appetites, or the enslavement to the human appetencies

of this local, planetary existence. I. i. 35–37 ; ii. 2 ; iii. 1.
When man follows the impulses of these violent passions
or fanaticisms, he certainly does as he wishes, but this is
not Moral Freedom ; but when he conquers these passions and appetencies in the Love of a higher life, he also
follows his wishes, but they are of a different character,
and this is Moral Freedom. When a man subjugates
the temptations of these animalistic and human desires,
and acts in the clear consciousness of the rightful purity,
he is seen as a Freed-Man, in the highest form of spirituality. I. iii. 29. Life is a struggle, a discipline, an instruction, and an education for Moral Freedom, yet
always in the order of the progression of the Almighty ;
and his perfection consists in discharging the duties of
*the moral life assigned to him in the time and place and
environment of his own and the world's progress. If
he is called to be a father, let him be a moral father ;
if a son, let him submit in moral obedience ; if a priest,
let him be so in gentleness, humility, and piety ; if a
servant, let him sanctify his life ; if a citizen, let him
yield to the Cæsar all that is morally indifferent, but
in the whole, in all things, let him strive for the perfection of the Sense of his Responsibility, for in it is the
sanctitude of his Moral Freedom. Man must learn, must
love, and must actuate his knowledge and his love in the
obligations of his spiritual nature and not in the violence
of his animalistic and human desires. II. vi. vii. viii.

8. In the historical ministration of Moses, the ideated
God and the proleptic morality are given in the concise,
but complemental system of the Decalogue. It is the
synthesis of duties bound together in the highest con-
ceptualization of the Love. It correlates the passional,
intellective, and affectional elements in man, and subordinates the whole. It is the concentration of the whole

sum of the movement for the moral life of Humanity.
It contains what is essential in all generations, epochs,
and localities of the world for presenting and preserving
the purity of man in and under this morality and the
awe and worship of the Almighty God who made man
of the dust of the earth, and gave him an Inner Eye
to see Truth in God, and draw and bind him by the
chains of a loving obedience to the discharge of duties.
But now, as then and in all ages, and in the ages yet to
come, large masses of the children of this proleptic and
educative movement have not reached, certainly do not
possess the mentalization for grasping this proleptic truth,
and for perfecting in their minds an image and likeness
of God ; and to restrain, instruct, and educate these, the
wise ministrations of this age must provide its system of
forms and ceremonies, yet free and flowing, and such as
are adaptive to the current of the ages. I. vi. 47. Around
the ancient Decalogue, always old as the pre-creative
ideas, and always new as the human motives and intentions which are to be governed and directed and elevated,
was built up that wonderful but temporary system of
ceremonial forms and severely punitive laws, so strangely,
and, for that age, superhumanly adapted to those early
times of undeveloped and unnormalated life of the tribes,
by which to conserve, unfold, and inwork into the life
of Humanity the everlasting obedience to the complementary correlations of the Law, which binds all together. II. viii. The Law has its perduring sanction
as long as man is in this theatre of life in his present
organization ; the temporary system had its temporary
obligation only as it conduced to the knowledge and growth
into the moral life of humanity of the former ; the former
subsists, the latter historically remains as the scaffolding which aided to build up the ideation of God and of

Law in the life of Humanity, and make it the great temple of living truth. Obedience to some of the many forms which the latter assumes in the currents of the ages, in the house, the farm, the school, the shop, the street, the forum, the church, is essential as a process for the attainment and conservation of the former. But these are only *hindrances*, except as they are taken as expansive forms through which the introduction of the former may be commenced and consummated. Obedience to God, who formed man with his correlations to the divine Self, requires submission and moral obedience to the processes for attaining the knowledge and the love, and inworking them into ourselves and into others, in the full comprehension of that law as a life — as the life-forces of the proper Self, as the means of human progress and the final atonement — καταλλαγην, I. vi. 44, of Humanity. If any child of vice and sorrow has any ideate, however low, of God, it is not to be destroyed, it is to be taken and expanded — unfolded. This only is growth, is learning, education.

9. If there ever was a human creature which fulfilled — was obedient to the Law, in this its highest sense, he was perfect, and no atonement was necessary. " He that keepeth the law bringeth offerings enough." " Love is the fulfilling of the law." To the extent of disobedience is conciliation necessary. To the extent of the self-imperfection, to the extent of the animalistic strength of the human autonomy — the " will of the flesh" which oversways the proper Self, and to the extent of the imperfections and perturbations of the psychical orgasms which give force and preponderance to human desires, ambitions, pride, vanity, mesmerisms, and fanaticisms — the " will of man" and the consequent obscuration of the spiritual solidarity, will an ascending normalation —

atonement — be necessary for the conciliation to the "Will of God." I. vi. 44. Enoch, Elijah, Christ. And it will be seen in the complexures of tribal and historical life, that in the more cultured nations, and in the more highly organized and mentalized individuals, there will be those who grasp these ideations readily, and by a self-imposed obedience conquer and subjugate the animal-man ; and that many require the temporary frameworks and the discipline of human laws and social opinion and coercive restraints to bring them slowly to their cognition and observance; and that multitudes are beneath all such influences. As, therefore, obedience in love — in moral freedom — is the fulfilling of the law, so, when this obedience is given, the law is perfected and the atonement is accomplished, or, in other words, no breach of the law, no atonement. The conciliation between the zoic life in man and the life of God are at one. The imperfection, the depravity in man consists in the low standpoint which he occupies beneath the knowledge of the law in its vast range of correlations, touching as it does the highest and the lowest extremes of humanity, and in the want of that love which works meekly and calmly in the actualizations of life in obedience to that divine prolepsis which requires him to ascend toward the deific standpoint in the everlasting immutabilities of Wisdom and Love, and in this living and ascending life to radiate light and love into the dark valleys and deep chasms beneath him.

10. The Prolepsis moves on ; and in the centuries, in each century, brings the billions of human agents, teeming in every latitude and longitude of the habitable earth, in varied aspects of autonomic organization and moral condition. In moral life, as in nature, God acts by secondary causes and intervening agencies ; yet he will be Governor in virtue of the prearranged prolepsis which, as Creator,

is the necessarily instituted foreplan of his Wisdom and
Love, but with contingencies for ever-changing vicissitudes and recurring presentations to these multitudinous
numbers of ·discipline and instruction and educative
growth in the manifold forms of individual, domestic,
social, political, and religious strifes and harmonies. Amid
these strifes and harmonies the sense of moral obligation
slowly unfolds. Conscience begins its work. And in the
more or less complex vicissitudes of life and character
it settles into some compromises with the animalistic and
human passions and appetites, and presents in the spiritual
life the same distinguishing characteristics which the
natural laws present in pygmies, dwarfs, deformities, and
monsters ; and as these have more or less likeness to the
human form in its nobility and majesty, so those other
present their spiritual similitude. But few, few reach
that full birth, which comes sometimes, with those throes
of agony which the mother feels in the birth of her first
child ; I. iv. 24 ; vi. 18, 40 ; and yet again to be repeated
in the ever-renewed contest with the fleshly and the
human desires, and as these prevail to sink down into the
form of man or the animal, or if the Spirit shall conquer,
to rise into the majesty of a sublime simplicity. Throughout the whole of the movement there is a necessity for
the subordinate processes, as the rungs to the ladder
which rests on earth and its top in the heavens. These
subordinate processes are the means by which submission
and obedience are secured in the preliminary and advancing gradations by which the Self escapes from the
dominion of the animalistic impulsions, and the human
and earthly loves to the normalation of its higher life,
and which strip the solidarity of all that is earthly and
human, and bring it into concordant harmonies with the
divine perfection. I. vi. 33. And thus it is that " if any

man build on this foundation, gold, silver, precious stones, wood, hay, stubble; every man's work shall be made manifest : for the day shall declare it, because it shall be revealed by fire ; and the fire shall try every man's work, of what sort it is ;" " for the mercy of man is toward his neighbor, but the mercy of the Lord is upon all flesh ; he reproveth and nurtureth and teacheth and bringeth again ; He hath mercy on them that receive discipline, and that diligently seek after his judgments." 1 Cor. iii. ; Eccus. xviii.

11. In this ascent the importance of the subordinate ceremonial and disciplinary processes securing submission unfolding into obedience are seen. The fear of force, of wants, and of human discipline in human laws and actions, is the beginning of human prudence, and the fear of the Lord in the discipline of the natural, physical laws, and the effects of indulgences in the animalistic and the human orgasmic forces bringing their causal punishments inwoven in the organization of man, and the proleptic providences in the history of nations, is the beginning of divine wisdom in man. Here is seen the double aspect of discipline and obedience. There are some who grasp the higher law of obedience, and others only the artful cunning of human prudential obedience, and others again only the " rule of conduct " ingrained in the animalistic impulsions. And this gradates civilized societies as it gradates the tribes of men in their savage, barbaric, and semi-civilized conditions. The spiritualized man has his law of obedience, which, at once, is seen as other than the cunning, artful prudence ; as his law, in turn, is different from the animalistic man below him ; but the grades imperceptibly mix in deepening and darkening colors or clearer lights from one into the others. In the whole of the ascent there is an escape from lower to higher laws.

I. vi. 41, 44.   As in the Mosaic Ministration some had
the law of obedience in their own clearer ideations, but
in their double face looking still to perfections above
them and to degradations beneath them, yet higher ascent
was impossible for them without duties of elevation to
those beneath.   It is now not only the law of spiritual
growth for the superior, but it is the law of temporal
security against the ruin and desolation of nations in the
multitudes which otherwise will fill and throng and seethe
and ferment in their deep corruptions; while under the
ceremonies and municipal institutions, in the effects of
successive ages, the Hebrews redacted, crystallized, as it
were, into a permanent race, yet did not sink into the
degradation and corruption of the Assyrian, the Syrian,
• the Egyptian, the Greek, or the Roman.   They still conserve their moral identity.   Their system of temporary
obligations educated their race to a point far above the
synchronic tribes of that time.   Through them a complemental system of law, correspondent to the Personality
of the Godhood and to the correlations of man to him and
of man to man, was thrown into the current of historical
human life.   The Wisdom of this Law was affirmed in
the Wisdom — Δογος — which appeared at Jerusalem,
yet by the unfolding of a higher form of the Moral
Wisdom, II. viii., and its struggle in and with philosophic
thought and human passions and affections for eighteen
centuries has been the unfolding of its great fulness
reaching to all the moral wants of humanity; and the
fluctuations of governments in their ever-changing vicissitudes of forms, dynasties, prevailing opinions, and philosophies, in their incapacity for making provisions for
moral obedience in the Superiors and a foundational submission in the multitude which shall be a moral system
leading directly, in its disciplinary education, into moral

obedience, has but strengthened the intellectual and moral foundations of that Law, and built its superstructure in the fabric of society — yet, however, rude and unfinished and discolored and stained with the blood of governmental and ecclesiastical victims and the corruptions of the race. Dynasties appear and disappear ; kingdoms and empires melt like brass in the furnace only to be recast into new forms, and grow cold and harden into despotisms ; republics pass away in a tempest of human passions and corrupt desires, and the armament of the battle-field is the iron foundation of the future empire ; and there is no peace nor hope for man, guilty or guiltless in these disasters, save in the consciousness of his spiritual life, and in this consciousness, in harmony, with others, conserving the Truth as the seed-time of a future harvest. When not trained by ceremonies and forms as the instrumentalities of peace and charity, these sad vicissitudes are the school-house of humanity, and the memory of victims may make the remorse which shall pray for mercy, and sorrows for a ruined land which shall kindle fresh sympathies in every age. The cultured Few, the appointed or provided guardians of Moral Life, from generation to generation, can only use their advanced position under the direction of the ideations and life of love involved in the move-ment, — in the peace and love and gentleness of that suasive intonement given by Wisdom and the holy Spirit of Love, and which in the dissolutions of the ancient governments and societies provided the elements of a continual re-formation for the ages coming and to come. It is their duty to evangelize society in forms instinct with the divine intelligibilities, and avoid the fraud and force of human instrumentalities, and not permit their holy offices to become *secularized*, for the spirit then sinks down to the man or the beast. And woe to that

people whose pastoral guides and spiritual leaders shall
secularize their hearts and minds with time-expediencies
and governmental fanaticisms, instead of intoning them
to deeds of gentle love and peaceful duties, and who
shall play the panders in the strifes of elections, and thus
conciliate the selfishness of the mart and the exchange,
and propitiate the demon of Secularization they may have
contributed to arouse, and which can only be exorcised
by streams — many streams of human blood, from the
avengers and the victims. Woe, when the Priest, of
any sect, fatally secularizes this multitudinous mob and
debases its motives of action to its human purposes and
objects, and instructs it to deeds of violence, that good
.may come, for its unevangelized purposes and passions
are as a two-edged sword cleaving both ways, the teacher
and the taught, and sharp with the jealous wrath of ·
Almighty God for the vindication of his own laws as
given in the ministrations of his Wisdom and Love.
Woe, when the hand of fanaticism and the weapons of
the unsanctified orders are red with blood; the axe of
the executioner is sharpened, the office of the hangman
is dignified, the bigot's iron rule of narrow thought is the
measure of conscience, the fagot is given to the incen-
diary, the most cruel monsters are the most highly hon-
ored, and in their guilty glory and vicious praises the
mob sinks deeper in corruption ; society is, for the time,
dissolved, and rape and plunder and unhallowed desolations are the work of those fierce multitudes which rise
up in all revolutions and follow their temptations and
indulge their appetites ; and honor and virtue and hos-
pitality and piety have no security, and " the blood of
the people is poured out as the dust, and their flesh as
dung ; " — but glory and grace light up the world when the
enthusiasm of the Divine Love inspires. It is Cæsar —
or it is Christ.

12. Man is placed in life in such conditions that he can only develop and there normalate his life. This he does as governing and as governed, from the child at the fireside to the boy in school, the apprentice in the shop, the servant in the Sabbath of his master, the workman at the bench and in the field, the citizen at the mart, the forum, the election, and the priest at the altar-service, and this from the elements within and around himself. In his early life and in his lower tribal forms he can only know the law by its infraction or by the tendencies within himself to commit the breach, and by the superimposition upon him of a disciplinary education, by subordinating instruction in coercive restraints and mandatory duties, by a ruler or power having a higher normalation of the law into his life. I. vi. 35–39. He cannot comprehend the higher forms of law and the life of the law ; but, as the autopsic Self becomes self-conscious in its movement upward, and clears up from the disciplines and instructions and education of the progress, the consciousness of want of conformity of the individual life to the ascending higher life, always just above it, enters, and the Self begins its ultroneous movements into its lower or its higher gratifications, and sin is consciously accumulated, or a depuration — a draining off of the dregs — a purification is exalting humanity. Involution in this maze is the normal condition of man as an individual and in his tribal divisions. In the processes which intervene between the imperfection — the depravity of the race, and the clearing up of his ideative intusception into the meek and unselfish love, there is an ever-recurring miracle, or there is a greater or less lengthened term of suffering, discipline and education ; — and even with the former the latter is a constant accompaniment. And this is his evolution. I. vi. 41, 44, 48, 49. The blind, actuous, objectifying

power acts very much as a spontaneity, although in the
conscious presence of the Self, from the stimulus of the
animalistic and the human appetencies ; the intellectivity
opens slowly, and light as knowledge comes by degrees,
and the order and harmony of life, commenced among
the wild nomades and hunter tribes by the observance
of forms and ceremonies, and in modern life by the complexures of civilization in its governments, wants, necessities, and comforts and sciences, its philosophic thought
and its religious dogmas, lead, in the main, to ideative
views which unfold into clearer vision of the higher life,
and to the knowledge of the superintendency of the
Divine Master of all life. The Love, wrenched by
sufferings and the mutations and trials and disappointments, in the destruction and change of earthly objects,
and the fruitlessness of human successes, seeks God in
the ideative fulness of his power and excellency, and
finds Him coördinated in his glorious Coessentialities to
man, through his ever-widening correlations, building up
his moral life.   In this ascent, in tribal, historical, and in
individual life there is an ever-expanding Law which
unerringly seizes him at every step and places him on
his election for goodness or crime towards himself and
the multitudes around and beneath him, yet in such way
that he can and cannot give, and they can only receive
as they openly aspire and win.

13. As in any mechanism there are difficulties and
dangers which are unremovable, as in the natural arrangement of hill and valley and watercourse incon-
veniences and injuries may arise, as in the production of
fire which may burn, and water, essential to life, which
may drown, as in any system of human thought, in some
direction, contradictories must evolve, as in the conduct of
life there must be conflict and suffering even for moral

success, as in forms of government the form, which in
repressing the lawlessness of the many gives too much
power to the few, and which in giving power to the few
takes too much of liberty from the many and robs them
of the means of comfort and moral progress, and which
cannot be obviated by any change of form or modification
of the system, — so in the economies of the divine prolepsis towards man there are disturbing influences in the
moral system which arise out of the imperfection, the
depravity in man, however derived or why permitted.
Accept the positive fact. This is the *Vice* of the systems
both physical and psychical. Distinguish the thought
from all previous use of the word as implying Sin.
Human life in all its multitudes is therefore *vicious* in
its want of knowledge and love, and these it can get only
in a normalative life. Such was the life of Paul, and
in his humanity " the Captain of their salvation was
made perfect (τελειωθείς) through sufferings." II. viii.
Hence the intermediary processes for attaining this
knowledge and realizing it fully in the Consciousness and
giving it a living intusception are found in the temporary
and changing vicissitudes of life, as child, man, parent,
servant, workman, neighbor, citizen, and saint, each with
its ever-changing and varying lessons in the education of
life. Thus man gets the appreciation of the higher life.

14. Evil — evils are those results of any causes which
produce suffering or prolonged inconvenience to man as
the mediate or proximate effects of such causes; *e. g.*, a
tree may fall and injure or kill, or a beast may kick or
gore ; or it may be produced by a free agent in the
proper exercise of his appropriate functions without any
sin on his part. Evil is therefore not properly synonymous
with Sin, nor always with the effects or consequences of
sin ; for, in the adjustments of God, sin may produce

good ; — "it must needs be that offences come" that
God's good may come. Evil always includes the consequences of.sin, which are hurtful and injurious, whether
to the one who sins or to one who is injured by the Sin.
Even subjective evil in its intrinsic fact, as in idiocy,
congenital monomania, is but synonymous with vice in
the guiltless imperfection of the organization. Evil may
therefore be the result of sin or vice, or of the working
of moral causes, which is but a further illustration of the
term vice. The jury or judge who condemn a man to a
heavy fine may take away the sustenance or means of
education of the innocent.

15. Sin is therefore the conscious violation or non-
performance in thought or word or deed, or where these
are proper, in all of them, of anything which is profitable
to the attainment and conservation of that state of the
Body, Soul, and Spirit in which the actuating power
objectifies its thoughts, words, and deeds into the currents
of its own and others' lives and is subordinated to the
intellectivity, and in which the intellectivity is expanded
and normalated into the knowledge and actuation of life
under the suasive attractions of that Love which was the
Causation in the beginning and is the Causative-end in
the end. In the aggregate of this complex movement,
as it must be caught by any full synthesis of thought, or
as the analyzation of individual experience or historical
revolutions and its power for the rejuvenescence of
society unfolds and brings it to light, it will dawn on the
mind that this Law is of the nature and essence of God,
and that it is the movement in Deity of his coördinate
powers in their coessential equality. Man in striving to
this gains his likeness to God, his obedience to law.
It is forever the Law seizing the individual and placing
him on his election ; yet in the large range of the move-

ment upward it is the election of submission to authority, and its end is discipline and instruction in ceremonial observances, distinct commands, punitive law-forces in their causal consequences, until in the processes of mentalization in individuals and tribal types, it ripens into the genial and loving obedience of that Law which is Power and Wisdom and Love in coördination.

16. In a system of proleptic, unfolding moral order, Subordination is neither the Vice, the Evil, nor the Sin of the System. *It is the system itself* of the All-Mighty, All-Wise, All-Loving Divinity. Other resultants may be the vice or the evil of such system, and Sin may be, nay indeed is, the conscious abuse or misuse, by ruler or ruled, of the subordination which is the very life-idea of a progressive moral order. Moral Progress is the Intermediate, and Moral Freedom is the Final Cause, I. i: 15, 42, 43, of an unfolding progressive moral order — the Prolepsis of the Almighty. Vice and Evil and Sin are hinged on and in the contingencies of Nature in its physical causes, I. ii. 19 ; iii. 3–8 ; iv. 13 ; vi. 2, and of Life in its instinctive and psychical causes, and in the Autopsic action of the Spiritual Self out of which this progressive moral order, this Education of Humanity, has heretofore and has yet to arise and advance. Government, in some of its historical forms, Subordination, in some or all of its manifold forms of domestic, social, political, and moral obedience, shape the advancing gradations from the savage and barbaric conditions through to the full, normal, and complete Emancipation of the Solidarity of the Race. I. i. 34, 33, 31.

17. It is seen that there are momenta to action incorporated into the organization of man, similar if not consubstantial in kinds to those inwoven into the animal organizations, such as the appetites of the stomach, the

venery, instinctive self-defence, &c. It is as fully apparent that man as man also possesses passional and appetizing momenta to action in various directions, and towards various pursuits in life, and that these are more or less variously complexed with or in higher or lower forms of organic intellectivity, as in the natural *artistes* and geniuses of different kinds, which impel to action with great uniformity, not only throughout the life of the individual, but in many of their forms as the general characteristics of the race. Many of these differentiate powers, thus inwoven in the human economy, come clearly out to view as the orgasmic propensities common to this human nature ; and that such is the case is seen in the spontaneity and separate action of organisms, as in dreams, monomanias, reveries, senile insanities, &c. ; and in their visible and sensational effects on different viscera of the organization, as shame on the cheek, awe on the scalp, anger on the chest, fear on the lower muscles, bowels of mercies, &c., &c. ; and in their control, modification, and repression or intensification from the autopsic direction and concentration or withholding of forces by the Self, — which latter is only a concentration of forces in some other direction. These orgasmic forces, animalistic and human, are continually in the life of the individual and the tribes impelling and appetizing to action and gratification. I. vi. 32. These are the animalistic and psychical momenta to action, and they necessarily imply diversity and locality of organs in a system of differentiate organisms for the specific manifestation of their separate functions, and these for their abiding inherence, intercorrelations and diversities of action, reaction, and interaction, and for their subordination to the unitary, the systematic control by the conscious autopsic Self. I. v. 1–14, 31–35 ; vi. 2–5, 16–20,

39-44, 46. Such an organization, composed of so many separate functions for building and enlifing the various parts of the somatic structure, so many of animalistic instincts, so many of human passions and appetencies, and all mediately and immediately under the control and direction of the autopsic Self, makes it a necessity of thought to transcendalize the idea of a wise adaptation of differentiate functions and of special organisms for demonstrating their respective actions and the eminency of the conscious, autopsic Self. I. iii. 5–15 ; iv. 3, 4. In this complexure of interwoven and reciprocating organisms is the autopsic self-determinating Self. This Self begins at zero in its ignorance, the *tabula rasa*, or blank sheet of Locke, and its cognitions and its combinations of these cognitions into systems, I. vi. 25–28, are mainly, as is its time and place, in the tribal or historical movement in which it appears. In its time and place in the prolepsis it gathers its cognitions, its intuitions, its ideations — its religious opinions — faith. Around this Faith, such are the historical and daily observable facts, whether it is fetichism, obeeism in any of its forms of snake-worship, polytheism in any of its forms of superstition, with or without human sacrifices, or monotheism — around these ideative forms, low or higher up, these orgasmic passions and appetencies cluster and swelter and appetize and impel. Are the Forms of Faith, (whether derived originally or they are, in the processes of life and education, impressed by all those means which contribute to the formation of habits of thought and action,) the doctrines and ceremonial forms of Mohammedism as thus impressed upon the individual and the tribe in the time and place of their historical movement, then, these orgasmic powers give vitality, powers, forces, actuation, and Mohammedism is actualized

by and through them — thus is it demonstrated into life
as the actual, practical religion of such people. Without
this vitality, derived from these passions and affections, it
is only a dead faith; and as a living faith it actually
represents the doctrines and emotional qualities which
so together make the faith. This faith, of whatever
particular form, when inwoven into the habits of thought
and action, moulds and intensifies those orgasmic powers
which are most naturally allied and correlated to its doctrines. If it teaches fatalism, its followers are bold and
reckless; if human sacrifices, they are bloody and remorseless; if the worship of Aphrodite, they are lascivious and
voluptuous, — so is man a stoic, a cynic, an epicurean,
and the natural character is intensified by the doctrine.
So the actual life corresponds to Buddhism, Brahminism,
Obeeism, Fetichism; and when Christianity is taught as
a religion of vengeance or strife, the terrible orgasms
inwoven in the organization of man, ever prompt for action by their very spontaneities, are brought into play, and
give frenzy and virulence to the internecine slaughters
and persecutions of civil and religious war, while a higher
culture of life in a purer doctrine gives the patience, the
charity, and the self-sacrifice of the Moral Logic controlling and moulding these orgasmic forces to higher
uses. I. iii. 29, 5–15; vi. 24, 40, 47. Whatever the
Faith may be, it becomes by the growth and habit of life
inwoven in the very fibres of the concrete existence. It
is not a mere abstract ideality. It is a concrete growth.
The child of genial organization in a well-ordered family
loves his parents and brothers and sisters tenderly. This
is so uniform that it seems as if nature had not only
implanted this aptitude or necessity to love, but that it
gave an instinctive knowledge of the domestic relations,
and as so it received the name of *storge*. Franklin proved

the fact to be otherwise, as but slight reasoning would have suspected or determined; and the clear distinction between the intellective cognition and the affection, instinct, or feeling is apparent, and the same child of genial organization, if raised from infancy among kind strangers as their child, would feel the same affection for its foster-parents — all things equal. If reclaimed by the parents at puberty, and compelled to leave its happy and genial home, where so many *attractions* in so many forms of love, I. v. 33, had grown and strengthened in his nature, these affections would be lacerated, and the metaphor, of nerves torn up by the roots, would be realized. So it is in the progress of life or the movements of civilization, when individuals or tribes move onward from lower to higher conditionings — from a lower to a higher Faith, — when they advance from the animalistic to the human, from the human to the spiritualistic, — from the obscuration of individual or tribal infancy to the clear veracious light of spiritual manhood. So man loves the lowest forms of Faith in which his infancy has been nurtured, and the struggle to rise to higher Truth is the laceration and breaking up of these habits of thought and feeling and actuation. I. vi. 31. So when he sinks — falls from his better estate. It is in and out of these conditionings that the Sense of Responsibility unfolds, gathering in its advancement knowledge and love and actuating power in higher forms of moral life. This Sense of Responsibility reduced to its elemental components embraces a knowledge, Cognition of the Right, Love of the Right, and objectifying power to Actuate the Right. *Ante,* § 7; II. vi. vii. viii. The Self is in this vast complexure of multiplied and various organisms, and these organic powers are seen as constantly improving or deteriorating under the diverse kinds of life demonstrating in and through

them. They are seen as the mere instrumentalities of
the animalistic or of the human life, or of the spiritual
life, or complexed of all of them; and they constantly
correspond to the life so demonstrated — actualized.
Thus the Faith of the individual, tribe, or people is manifested and presented in actual life. When these orgasmic forces are exacerbated, and are acting with the
violence and fury of a popular madness, what voice of
reason or charity can be heard?

There is an action from all causations in nature and
life, which environ it, on the Self, and this Self reacts on
nature and life. In this reciprocity of action and reaction,
the Self, in its capacity and modes of acting in its animalistic and human organisms, is altered and changed. I. vi.
• 1–3, 17, 82; iii. 5–15. The Alterability of the human and
even of the animal organizations by the action and the
direction given to these orgasmic forces as they move
into action in a mere state of nature, or. as they are
modified in the forms and uses of civilization and of
religious faith, higher up or lower down, and this modification of the organization of man as the movement of his
conscious aspiration after goodness and purity in self-
control and self-direction, which have, so frequently and
palpably, been presented at various turns of this unfolding system, tend constantly to mould the individual, tribe,
and nation in bodily, psychical, and spiritual conformity
to the predominate direction given to the tribal or national
life, yet subordinate to the final education of the race,
and to be inwoven into its general purpose and end in
the progressive fulfilment of the deific prolepsis. I. iv.
25; v. 14, 20. Thus again is presented the importance
of cultivating the ideative processes in the education of
the race for the direction and control of these orgasms,
and so for the disenvelopment and consequent depuration

of the proper Self. I. i. 1–4 ; v. 26 ; iii. 5–15 ; ii. 13–15 ; iv. 18–20, 30 ; v. 8–14, 31–35 ; vi. 39, 40, 44, 46. When the true Moral Logic is not truly normalated for the subjection and direction of the passional and affective spontaneities, and these, instead of being restrained and directed by the Self to the consummation of the order, the justice, righteousness, of the All-mighty, All-wise, All-loving God, are or shall be misdirected by perversions and false teachings of Faith and Practice, and intensified by the infusion of the conscious, the autopsic powers of the Self into these fatal orgasms, the end must be intellectual and moral confusions and desolation among the peoples. It is thus that manias, monomanias of individual life, of endemic insanities, and national and ecclesiastical fanaticisms, are so frequently engendered and produced. I. v. 9–14, 22–25 ; vi. 17–20, 28–32. Thus the tiger and the monkey in their organic characteristics reappear in the human successions, and murder and mow and chatter in the crowds and processions of life, and make history the record of their passions and their persiflage in the desolations and desecrations of Humanity. I. ii. 12–15. So to the observant eye the animal characteristics appear in various forms in society, — the crow, the beaver, the mocking-bird, the leopard, the lion, the fox, the badger, &c. ; and it is only by the supra-tending action of the spiritual Self that it reaches to knowledge and Love, and a power of Actuation of such self-actualization, and assimilates more and more to Deity. In animal nature, and in the animalistic orgasms in man, these organs act in virtue of the law-forces inwoven and differentiate in them, giving them their special forms and functions ; this is their "rule of action." I. i. 30 ; ante, §§ 2, 3 ; iv. 3 ; vi. 37, 44. So in the functions of the merely psychical life in man the organisms act in obedience to the *un*consciously intelligential yet differentiate forces inwrought into their respective functional activities — and this is their " rule of action." They act directly as cause and effect, so far as they act or are permitted to act without control by the self; and clearly so when they overmaster the Self. I. vi. 44. But when the Self acts in its own conscious autopsic independency, it does not act as blind cause — as law-force producing designate and invariable sequences as effects, but it is consciously autopsic Self-Cause determining and producing effects, I. iv. 5–16, and the whole moral nature of man lights up and unfolds into its spiritual presidency, controlling, subordinating, or in lower or higher forms of moral life moulding these animalistic and human law-forces, and reaching forth, consciously and aspiringly, to its moral destiny in the knowledge and love of God, and in the meekness and gentle firmness of this love to the actuation of its duties. All along this line of progress the distinctions between animalistic and human functions, yet in the consubstantiality of their underlying forces and the clear consciousness of the autopsic self-cause in man, come out to view, and in the gradations as distinguishing individuals, tribes and nations and peoples, which all the causes at work in and on and around the organization of man concur to produce, the vice of nature and of this composite life is seen as inherent in the system of things, and Evil flows alike from natural, instinctive, and psychical law-forces of their various kinds, and their necessary conflicts or different forms of power and action, and from conscious sinful cause warring against the purity of this spiritual Self-Cause in man. These causes gradate society, and the order of the Almighty is an order of proleptic progress which imposes at every step in the procession of Humanity the duties of the Teachers and the lessons of the Learners. Do not mistake the half-taught and less than half-learned instruction of words — words which become cant, for the ministrations of duties in actual, positive life. I. vi. 32, 39, 46–48; vi. 1–3, 22, 40. Thus it is seen that all thinking, loving, actuating, which have not impressed upon them the love and the service* of Almighty God in the constant presence and control of the Moral Logic, are purely animalistic or purely human, or complexed of both of these, and therefore have no element of love or conscious service of God, and are of the earth, earthy. To comprehend these things, man must go into the depths of his own nature and understand Saint Paul when he said, " For what man knoweth the things of man, save the spirit of man which is in him?" I. i. 1–3; iv. 29–31.

18. To take a more metaphysical view of Evil and Sin, and avoid the moral contradictories and reconcile the historical and geologic facts involved and the general hypothesis of theologians that physical evils, which long preceded man, were the result of man's sin, authority and reason may give a conciliation; — " The Church has never defined the duration of the period of time which elapsed between the creation of the first elements of the world and their coördination on earth and in the heavens," and it has never " defined that the days of the Mosaic Cosmogony were days of twenty-four hours." — *Prot. & Inf.* c. iv. § 4. " Did Christ come to teach men the arts of commerce, to render them skilful money-makers, to train them in the construction of railroads, steamboats, and cotton-factories?" — *Id.* c. iii. § 2. Evil must be cognized as a metaphysical *accidence*, I. i. 6, and as resulting from good intentions or bad intentions, or from no intentions and from physical causes. Evil is an effect of something else. It is not in itself a positive essence;

while sin is positively *subjective*, and involves the consciousness of guilt, and although not an essence, but an
accidence of the subjective Agent, it may obdure as long
as the sinful agent subsists, in his sinful disposition. Sin
is the accidence of the Spirit, or else it is essential to it,
and therefore sin as an ontology is eternal, or else it was
inwoven in its creation and so the direct act of Deity. It
only supervenes in the progress of life. Offence must
come that God's good may come. It is therefore an
accidence and capable of removal or remedy. That evil,
or those things which are capable of producing it, are the
direct work of God, may be affirmed upon these intelligible grounds and upon Scriptural authority, — " I form
the light, and create darkness ; I make peace, and create
evil : I the Lord do all these things," — Is. xlv. 7 ; Am.
iii. 6, and destructive monsters and venomous reptiles and
destroying agencies of nature accompany the geologic
eras long before man appears on the earth, and at the
time of his appearance. The Evils of nature, the
" disorder" as it is called, pervading the world, was not
then a consequence of man's prevarication or fall in any
sense of cause and effect, or in any sense of penalty for
a fault committed ; for these preceded and accompanied
the appearance of man on the earth — unless mankind
had a previous existence, and the earth was created as
his prison-house to probate and improve or punish him.
And this only removes the difficulty one step further back,
and increases the improbability of *any* solution. This
brings to view the broad distinction between Evil and
Sin. Evil, thus being a pure metaphysical accidence, may
arise directly from essential causes in the combinations
of secondary causes. Until secondary causes appear, there
can be no sin ; but sin, when it appears, is a subjective
personal fact, and it appears as conscious cause and

effect. I. vi. 44 ; vii. 18. This strikes at the root of all
Manicheism — a duality of Gods, one good and the other
bad, or that Justice and Goodness and Sin are in one God.
Evils therefore are not sinful, but acts which produce
them may be. Now ideating God as omniscient power
and omniscient intellectivity and omniscient love in their
coördination, (without which there is no God, I. iv.;
vi. 46,) neither Sin nor Evil can be predicated of him in
his essence ; the mind starting with these elements of
thought, cannot think it. But when the secondary causes
come into play as physical causes, psychical activities,
and conscious, limited moral agencies, and as they act
and react on each other and the last may consciously use
and abuse the others, nay, in the growth of life must
misuse them, evil and sin may, will both appear. The
mind cannot but think them as the necessary consequences
of the secondary causes. In a divine prolepsis, all of the
coördinations are present, and rule the plan of the movement. But in a prolepsis, a foreplan is implied of something to be done and consummated through successions of
eras, an imperfection and an intercurrence of evils to be
guarded, abated, or remedied, a vice in the system to be
removed or modified by the intervention of higher and
other causes, a sin as a necessary precedence to knowledge, and suffering as the condition of higher love ; and as
this conviction is unfolded, and this love becomes more
conscious and open in the growth and discipline from
infancy to age and from tribal degradation to tribal and
national responsibility, and evolves the noble grandeur of
Moral Freedom, the prolepsis will be seen as moving in
succession to attain the end of the succession in doing.
That which is at the end of the doing is the object to be
attained, the gratification, the love to be enjoyed in the
attainment. In a prolepsis the ultimate gratification is

at the end of the movement instaurated as causative-end,
and therefore the Love, in its richness and glory, is separated from the Intellectivity and the Power by the whole
series of causes and successions from the Beginning to the
End. I. i. 15 ; iv. 10–13 ; vi. 46. Man, then, placed in
the great web of these causes and their effects, and beginning in infancy as man and in a state of autonomic envelopment as tribes, must in and of and from himself, yet
in constant correlation with each other, and all depending
on the correlations with God, move forward through the
tribulations of vice and sin and evil to the *incarnation*
in themselves of the Knowledge and the Love and the
Actuation of Justice, as ordinated for the movement.

19. Justice is, therefore, that Law which seizes man
• at every step of life in his actual moral condition, and
subjects him to the discipline, instruction, and education
of the prolepsis for a still higher moral condition. Thus
all adversities may be sanctified discipline, all struggles
but instruction unfolding the Intellective Powers, and all
sorrows an education to that Love which would impose
no adversity, no struggle, and no sorrow, save in maintaining that order which best conserves to the moral
growth of life. This is the Education of God for
Humanity. God knows ; Man will learn. The same
application of the law, as a great horizontal line of
equality, to all the members of society, to all the tribes
and nations of the world, is folly, and it is madness. Each
can get but what they are capacitated, mentalized to
receive. It is a law of progress, and not a constant
miracle. It is not pantheism. I. vi. 47. As positive
fact, to those who only can see facts, but in the transcendental system, for those who can see God, there is an
arrangement of appropriate moral correlations between
man and man, and man and God, as inwrought in the

differences of the organic constitution given to each individual in society; and in their conditions in the world and in these very differences the moral correlations to God are necessitated on any system of thought which finds Power, Wisdom, and Love in any form in the universe. They are the coherences of society, they are the bond of union in state and church, they are the conflicts of all time; and their removal, in the conciliation of their elevation to the full spiritual disenvelopment from these Differences, is the Divine Concord. Man, in these differences, is placed in the complexures of this onward movement of life, and always around him, as active or passive, is woven the web of circumstances so as to prescribe to him the nature, the means, and the occasions, at every stage in his progress, for raising the questions of duty and promoting the growth and expansion of his mental faculties and his responsible election of conduct in the constant occurrence of identical or new presentation of higher problems of conduct on which his higher knowledge and his purer love shall be successionally evolved. The successional billions coming and to come can only thus be disciplined, instructed, and educated. The circle of these circumstances, so as to evolve the occasions of election and conduct, must be comprehensive as the race and minute as the distinctions between the individuals; for one will violate this law, and another that, and all in greater or less degrees of temptation, or as they are in greater or less advancement of moral life. Justice is, therefore, no fixed and absolute application of a fixed and absolute law to every individual of every tribe and tongue as standing on the same horizontal level of equality, instead of that inclined plane over which all are making the moral ascent or descent of life. It is a principle of administration in the various offices of life for each Self and towards others, changing with the condition and character of individuals, tribes, and nations. Reverence to some, regard to others, submission here, obedience there, authority now, and moral resistance to moral wrong everywhere, and charity, love for all. Such is the order of the movement. It is so found in the demonstration or on the concession of any God. It must be so in a God of Wisdom and Power and Love, forecasting the fact and the order of the creation, and providing for the discipline, instruction, and education of a race where there is an infancy of individuals and of tribes or races. It is the positive fact of life and of history. It is the Prolepsis. The child of tender years is not governed by the same fixed rules, either on the part of the parent or child, neither can he be nor ought he to be, as the child of maturer years; yet the parent, in obedience to his higher knowledge and the previsory love which provides for future growth and education of the child, must guard against his own inordinate human love, and he must educate him for a life in which he must act with and against others ; nor is the child arriving at puberty to be governed as the full-grown man living under the paternal roof ; yet throughout, the general spirit of the intercourse and the government should be the same — a proper admixture of Love and Power tempered to the age, qualities, character, and prospective duties in life. Nor are the same laws of action and conduct, though similar, applied or applicable to the individual in society and government which constitute the proper rule of Justice in the family. Yet there is a submission and obedience necessary in the family, an obedience necessary in society, and an obedience necessary in the government; yet they are all obedience, and each is different from the other, and each, in turn, is displaceable.

They are correlations, one of the other ; and where submission or obedience is *proper*, authority — power is
necessary. I. vi. 35, 36; *ante*, § 3. Yet always the
authority, the power must be founded in moral right.
You who have passed through these things and appreciate,
will understand them, whether they are further illustrated or not ; and others cannot appreciate them until
their mentalized lives can intuscept them and give to
them a life, a spiritual content. I. vi. 47. But in the
child, in the man, in society, in government, in church,
there may and must be an obedience, a submission
beyond the capacity to understand the occasion and the
reasons for the obedience, or else order is at an end.
But at this juncture, in the organizations and conditions
of individuals and of the races of men, occasions will,
and, in the very nature of the agencies at work, must
arise, when authority shall cease, when the parent
forfeits his moral right, and manhood must meet its own
responsibilities, when government shall violate the public
sense of justice too far, and tyranny in actual forms of
force and in destructive taxations sapping the moral life
shall become insupportable, as well as when society
shall, in its unnormalated and undisciplined numbers,
become corrupt, and the accumulations of viciousness
and crime shall result in immoral disobedience. But it
is seen that there is a justice which rules, yet changes
from beginning to end, from infancy to old age, and
from the lowest envelopment of tribal degradation to
the competency for a wise and rightful nationality of
self-government, which is but the growth in the very
order of the prolepsis ; and that, throughout, this Justice is but rightful obedience to the Master of Life.
Authority is to be exercised in a spirit of Love ; and
obedience is to be rendered in a loving, just, and dutiful
subordination. I. ii. 15.

20. In a state of things where perfect Order exists, the Actuating Power, the Intellectivity, and the Love are perfectly coördinated or correlated in their fulness. When the Actuating Power is weak, the plans, designs, the modes pointed out by the Intellectivity, and the means selected, cannot be used or made efficient; when the Intellectivity is circumscribed, to this extent the Intelligence of the agent is limited, and error in ends, modes, and means of doing is or may be introduced, and confusion, disorder, and evil prevail; when the love unconsciously, without a consciousness of higher duties, rests in the animalistic or human gratifications, the loves, then, for which this agent toils, is animalistic or human, and in the absence of a conscious knowledge of the Law is simply vicious; § 13; when the love consciously pursues the animalistic and human gratifications, then it is consciously human and animalistic, and is sinful, as in the former case it is vicious; but when the Love is divinely directed, then the whole of the powers, whether weak or circumscribed or enlarged, are turned towards Divinity, yet with an aspect towards nature and life, in virtue of the attractive sympathies of the Universal Love, and so for the purification and uplifting of all, the Self aspires and gathers knowledge, love, and power. Justice — Righteousness, then, is simply the Divine Order working in the confusions and disorders of humanity. Human justice — righteousness is, therefore, but the approximative attainment of human conduct for moving wisely and lovingly in the divine order for the purification of ourselves and the onward movement of the whole. It requires the regulation of our conduct towards others by which to maintain, and by maintaining attain still loftier heights of purity and divine love, yet in that humility and meekness which will return and act upon the raw and rude

elements of society so as not to involve the loss of any of the elements of our own purity, meekness, and humility, but so as to lead them through the ages by purified agencies to the Divine Order — Righteousness, Justice. §§ 8, 11. God is Order ; and Power, Wisdom, and Love, in coördination, is Order ; and a constant progress towards this, in unfolding these powers in the Solidaric Self and in the Communal Solidarity of Humanity, is the movement towards universal Order. § 16.

21. Such is the slow and solemn march of the great movement. Amid evil, vice, and sin, man must toil and move on to his perfection. Fraud and force are not *his* ministers, however constantly the Great Ruler may so ordinate these that good may come and make the wrath of man to praise him ; but he who uses them has stained the brightness of his glory, and written on indelible history, or registrated in the tablets of his own organization, the record of the violence of his passions. The great day of the thousand years will move on regardless of the utmost love of power, place, fame, wealth, or vengeance, however it may be exacerbated into fanaticism. The mighty fate of the Prolepsis seizes all, in some of its many forms of agencies at work — the animalcule, the lowest dwarf of humanity and its noblest genius, and each only works in their time and place in accordance with the causes around him, and no earthly power can lift any out of its intrinsic condition. We are akin to all nature, and all nature conspires against man to discipline and instruct him, and all nature is his friend and ally for his education and advancement. The lofty cannot subsist without the lowly, but the *low* will only toil and swelter in anarchy and confusion without the gracious and exalted. The scholar, the artisan, the · laborer, and the servant, of whatever form, are correla-

23

tives, and the position of each is as he sanctifies his position. Each can receive only the ministrations of this order, and this worthily only as he worthily wins it in the struggle to rise to a higher position conditioned in the moral correlations which he holds to all around him. Wealth, fame, power, in the possession of the unworthy, are but means and solicitations to greater abasement and more obduring evil. Divide property to-day, and to-morrow fraudful cunning will more than regain its share. The parent may accumulate, and the child shall waste. Destroy one tyrant, and many shall rise in his stead. Governments may change forms and dynasties, but humanity has but one lot — to work wisely and well in the order of the Almighty, and aspiringly unfold in wisdom and love, or, as the slaves of passions and misdirected affections, grope in confusions and battle in vain against those everlasting barriers of the divine order which close around him in his fierce conflicts and crush him in tribulations and sorrows which give no wisdom and no love. Force may gain power, and Fraud obtain unhallowed wealth; but the knowledge of the one shall perish or remain a memory of infamy in the desolation or silent despotism which always accompany it, and the moth and the worm shall eat the other from his clutches, or folly and guilt in vicious and criminal indulgences shall hold their carnival over his grave. Yet stand firm and look from your prison-house, with your fellow-victims who suffer or perish with you, and behold the Star above you. It rolls on forever, and its light has been a glory and a guide. You are in nature; and in nature all is cause and effect; I. ii. 19; vi. 2, 44; in animal passions all is cause and effect; in human passions, appetites, and affections there is vicious cause and effect, and in their conscious use they are criminal cause and effect as the instruments of violence and fraud, or they are the ministering agencies of Wisdom and Love. God is his own Avenger; I. ii. 19 ; iii. 29 ; and his deepest vengeance is sorrow for the misuse and perversion of the powers which should have been a blessing, but which have been used as a curse and a desolation. The Roman, the Goth, the Hun, and the Vandal — Titus, Alaric, Attila, and Genseric, were the ministers and the scourges of God ; but Jesus Christ and his followers were the ministers of Wisdom and Love and Moral Power which reconquered by suffering. The earth is planted with the blood-seeds, and they grow and ripen from age to age.

22. Man is in the web of the Prolepsis. And there is a vice in the whole system of nature and life. There is scarcely a fact, a *factum* in it which may not produce evil. His whole organization is vicious, and he escapes from the evils which the vice in the system of nature will constantly produce, only by the cultivation of his own intellective and moral powers. As he studies nature in his daily struggles with it, or in the pursuits of science, he learns how measurably to avoid the evils which the vice in the system of nature will constantly inflict. He gains the secrets of her powers and converts them to his use and misuse, and he or others learn by the misuse. He turns in upon himself, and in animal appetites and human passions and affections it is vice, with a constant tendency, in indulgence, to increase the vice and multiply the evils which flow from them and their increased solicitations by their indulgence. But man, standing in his place, in the early conditions of the opening prolepsis, imperfect in all his knowledge, confused, perplexed, bewildered, violently controlled by the passions and affections, or some predominating one in-

woven in his nature, and from which he, in his spiritual,
solidaric Self, is to be disenveloped by his own Conscious
Causation working in the ages, how shall he move forward to its attainment? I. iii. 29.   Certainly by his own
reason, as evolved in the processions, and as certainly
always in doubt, if not in despair, — yet certainly by his
own reason, — and for this it was necessary that "the
Law should enter that Sin might abound." I. ii. 19; vi.
34, 40, 35, 36, 44, 47, 46.   The consciousness of moral
offence is essential to an escape from the viciousness of
the natural condition; for without this conviction of
moral offence it is only a question of prudence in indulgences, and this so as merely to escape the natural evils
occasioned by the indulgence, and to make the best
compromise with the penalties involved, as all the moral
philosophies, ancient and modern, based on such views,
attest. I. vi. 44, 45, 47.   Without this consciousness
there is no inducing cause, no motive, and can be no intention to reach to a higher law or *cause* for conduct;
while Law as Command, and in its form as Command,
takes the inquiring Mind up at once to find the authority of
Command, and the foundation of Law is ascertained and
determined.   And this, in its processes and end, correlates man to God and to all between man and God.
Vice, evil, and sin are therefore necessary in a prolepsis
where man must make his ascent through symbols and
signs and forces in nature and life to the Supreme Powers
of all life. § 16.   Evils, which are the result of imperfection in a system which produces creatures without
any fault or causation in themselves, and which are by
their very constitution vicious, but yet are so constituted
that by various means of discipline, instruction, and education they are capable of progressing to a clear self-
cognition of the vice in their own natures, and must

thereafter normalate and unfold their own conduct to
guilt or innocency, are necessary. Such agents are not
the subjects of the same laws, in the same manner, in
the earlier as in the later condition. In the former, evil,
punishment is disciplinary and only can be; and in
the latter, it is punitory as well as disciplinary, for all
punishment in a wise government is disciplinary, and
God is wiser than man. The former cannot be the
object of punishment in like manner with that creature
which is consciously sinful or wilfully blind and dis-
obedient to the means of lifting himself above this sinful
ignorance and this vicious condition. Hence the moral
propriety that Command should lead to the vast compre-
hensⁱon of Law — of the Divine Ideas in their ever-
widening correlations. But neither the sinful nor the
vicious nature can be tolerated by the Divine Justice.
The consciously sinful nature is clearly at war, and in its
own consciousness, with the deific attributes of Wisdom
and Love which, in their coördination, is ideal, theoretical
Order; while the latter is also at war with both, but
unconsciously. With neither class can Deity have any
harmony; he can have no harmony with wrong-
headedness, for it mars and violates his Wisdom in his
divine arrangements; he can have no sympathy with
wrong-heartedness, for it violates his Love — in the processes and in the end of his divine arrangement. When
head and heart are both wrong from automatic or congenital causes, there can be neither harmony nor sympathy in this sense, for both are violated, and it may be,
not in the conscious sinfulness of the Self. In this latter case he can have no harmony of action, or of thought
or sympathy of Love; yet the Solidarity of the family,
the tribe, the race, are concerned in these correlations
which bind moral causes to moral effects and the natural

and psychical causes to their effects, and all into a system
of divine government, I. i. 32, 33, where Wisdom, Love,
and Power in actualization is practical order — justice,
always revealing enough to teach wisdom to the ignorant,
and always concealing enough to humble pride, and place
man on his meekness and humility, and to teach him
that Divine Justice is something other than the measures
of human vengeance, and, ever and forever, is inducing to
further knowledge and more chastened Love, still aspiring, until by the action and reaction of the forces in
nature and life, and the action and reaction of the forces
woven into and intrinsic to the Body, the Soul and the
Spirit returning to the creative concordance of the primal
image and likeness, "the day of everlasting brightness
shall dawn, and the shadows of figures shall pass away."

END OF BOOK FIRST.

THE
LIVING FORCES OF THE UNIVERSE,
THE TEMPLE AND THE WORSHIPPERS.

BY HON. GEO. W. THOMPSON,
OF WHEELING, WEST VA.

THIS work has received the commendation of some
of the first minds in America. Divines of the various
Schools of Theology have spoken of it in terms of
high praise, and its general circulation will be productive of great good in disseminating truths which,
if fairly considered and candidly applied, will tend to
change the materialistic doctrines and infidel radicalism which now so largely prevail to the injury of
the morals and religion of our people, and substitute
a conservative radicalism which will improve society,
conserve its institutions and promote progress. Its
method and its processes are entirely new. It offers
a new solution for the unity of the human race. It
solves the paradox of their intellectual and moral diversities, and lays the foundation for a new moral
science. It gives more definitiveness to the formulas
of scientific expression, and in establishing the underlying identities of the intellectual and moral power
of the human race, and their image and likeness to
the triune essentials of God it harmonizes the correlative unity of all science and literature.

It is the first decided step toward an American
philosophy.

Price $1.75, or 12 copies $15.00.

HOWARD CHALLEN, *Publisher.*

PHILADELPHIA.